Officer Swords of the German Navy

Officer Swords of the German Navy

1806-1945

Claus P. Stefanski & Dirk Stefanski

Schiffer Military History
Atglen, PA

Dedication

To our wives, Doris and Hildegard.

Photo and Object Credits

Photographs of Swords: (numbers of these pictures in the book are preceded by the letter W)
Army Museum, Brussels (AMB): 2, 18
German Blade Museum, Solingen (DKS): 63, 70, 71, 72, 73, 76, 81, 94, 122, 127
Prince Fürstenberg Museum, Donaueschingen (FFM): 171, 172, 173, 174
Historical Museum of Rendsburg (HMR): 33, 34, 35, 37, 38, 39
Thomas Johnson Collection (STJ): 67, 153, 154
Museum of German History, Berlin (MDG): 95, 96, 166
Private Collections (PS): 1, 3, 4, 5, 6, 7, 8, 9, 10, 11, 12, 13, 14, 15, 16, 17, 19, 20, 21, 22, 23abc, 24, 25, 26, 27, 28, 29, 30, 31, 32, 36, 42, 43, 44, 45ab, 46, 47, 48, 49, 50, 51, 52, 53, 54, 55, 56, 57, 58, 59, 60, 61, 62, 64, 65, 66, 68, 69, 74, 75, 77, 78, 79, 80, 82, 83, 84, 85, 86, 87, 88, 89, 93, 101, 102, 103, 104, 105, 106, 108, 109, 110, 113, 114, 115, 116, 117, 119, 120, 121, 124, 125, 126, 129, 130, 131, 132, 133, 134, 135, 136, 137, 138, 139, 140, 141, 142, 143, 144, 145, 146, 147, 148, 150, 151, 152, 155, 156, 157, 158, 160, 162, 163, 164, 165, 167, 168, 169, 170, 175, 176, 177, 178, 179, 180, 181, 182, 183, 184, 185, 186, 188, 189, 190, 191, 192, 193, 194, 195, 196, 197, 198, 199, 200, 201, 204, 205, 206, 208, 209. For reasons of security, no individual identification has been given for items from private collections.
Peter Tamm Collection (SPT): 40, 41, 107, 111, 112, 118, 123, 207
Military History Museum of Rastatt (WGM): 90, 91, 92, 97, 98, 99, 100, 159, 161
Personal Pictures and Facsimiles:
Federal Archives-Military Archives (BA-MA): 14
Prince Fürstenberg Archives, Donaueschingen (FFM): 34
Huss Archives: 26, 36, 45, 51, 56
Koehler-Müller Publishing Group: 3
Maritime Officer Organization Archives (MOV): 18, 24, 32
Maritime School Archives, Flensburg-Mürwik (MSM): 7, 8, 9, 10, 11, 15, 16, 21, 23, 27, 31, 37, 38, 42, 43, 46, 48, 49, 52
Rackelmann Archives: 40, 41
Claus P. Stefanski Archives (CS): 1, 2, 4, 5, 6, 12, 17, 29, 30, 35
Dirk Stefanski Archives (DS): 20, 22, 25, 28, 33, 44, 47, 50, 54, 55, 57, 58, 60, 61, 62, 63, 64, 65, 66, 67, 68, 69, 70
Peter Tamm Collection (SPT): 19, 39, 53, 59, 159
Military history Museum of Rastatt (WGM): 13

Book Design by Ian Robertson.

Library of Congress Catalog Number: 2002108063

Printed in China.
ISBN: 0-7643-1674-5

This book was originally published under the title,
Die langen Offizier-Seitenwaffen der deutschen Marinen
by E.S. Mittler & Sohn GmbH

Published by Schiffer Publishing Ltd.
4880 Lower Valley Road
Atglen, PA 19310
Phone: (610) 593-1777
FAX: (610) 593-2002
E-mail: Schifferbk@aol.com.
Visit our web site at: www.schifferbooks.com
Please write for a free catalog.
This book may be purchased from the publisher.
Please include $3.95 postage.
Try your bookstore first.

In Europe, Schiffer books are distributed by:
Bushwood Books
6 Marksbury Avenue
Kew Gardens
Surrey TW9 4JF
England
Phone: 44 (0) 20 8392-8585
FAX: 44 (0) 20 8392-9876
E-mail: Bushwd@aol.com.
Free postage in the UK. Europe: air mail at cost.
Try your bookstore first.

Contents

Introduction

Throughout history, maritime matters have been able to inspire great interest, and at times even fascination, in Germany. Today, for example, this is expressed in the widespread enthusiasm for water sports, interest in the nature of the sea, or—as in the case at hand—in the study of specific maritime themes.

In this book the authors take up a subject about which little has been published previously. Unlike the armament of ships, sidearms as equipment of officers and portepee underofficers in the various German navies have remained largely unknown. This may be in part because the navy of the Federal Republic of Germany, unlike the People's Navy of the German Democratic Republic's national armed forces, never introduced a sidearm for service or walking-out uniforms, and sabers and daggers vanished after the unification of the two German states in 1990. Yet a look at the navies of other countries, as well as the German land-based armed forces, shows a great deal of material on this subject. With this background, I am happy to greet the book at hand as a previously missing stone in the mosaic of German naval history.

On the basis of experience in the recent past it is quite understandable that many people have an inclination for everything that has to do with weapons. In the process it is often overlooked that not the weapon, but only the political or military purpose to which it was—and is—used can be judged by the opposing terms of "good" and "bad." In view of this, weapons must be regarded not only as instruments of warfare, but also as evidence of cultural history. In the past century, swords were still used by the German Navy for ceremonial purposes and to honor deserving people. They still have this use in other countries and their navies. Sabers, swords, and daggers are still presented by many nations as gifts to guests and high honors to deserving or befriended persons. As a rule it is not the weapon in and of itself, but the historical relevance and cultural value that matter to the nation.

It pleases me that two "shore rats," the brothers Claus and Dirk Stefanski, after long years of research in archives and examination of a great many public and private collections, can make the fruits of their labors as "scientific collectors" available to the maritime world, as well as those of the public who have an interest in military and maritime history. I wish the book a lively and well-deserved interest.

Hans Lüssow
Inspector of the Navy

Foreword

German naval swords have been examined closely in many past and present publications. The "Long Sidearm" of the Reich Fleet officers of 1848, the Reichs- and Kriegsmarine, and those of Brandenburg, Prussia, Schleswig-Holstein, and the Empire, though, have been given little attention to date. But this subject includes many a pleasant surprise. The 150th anniversary of the German Navy in 1998 inspired us to examine this subject more closely.

The work now offered to the public is intended to document the present state of knowledge of German naval officers' sidearms for the study of naval and weapon research, and to offer naval fans, army buffs, and sword collectors an overview of the long swords worn in the various German navies.

A project like this book cannot succeed without the help of friends, private collectors, museums, and other institutions, and at this point we offer them our thanks for their generously given help and support. Unfortunately, during the course of our research the willingness of official agencies here and there left something to be desired. The authors thus thank all those all the more who assisted their efforts (in alphabetical order):

Private individuals: His Highness Joachim Egon Prince von Fürstenberg, Mr. Georg Goerlipp, Mr. Edwin De Groote, Dr. Friedrich Hermann (+), Mr. Rolf Hofmann. Mr. Jürgen Huss, Eng. Thomas Huss, LTC (ret.) Thomas M. Johnson, Mr. Hartmut Kölling, Dr. Rainer Lackemann, Mr. Gerhard Lengelsen, Mr. Peter Meihs, Mr. Peter-Hans Rackelmann, Mr. Gerhard Seifert, Mr. Wolfgang Seybold. [+ = deceased]

Public Institutions: Dr. Ernst Aichner and Dr. Georg von Kern of the Bavarian Army Museum, Dr. Fuchs and staff of the Bavarian State Archives, Ms. Lorenz and staff of the Library of the Federal Defense Ministry, the staff of the Federal and Military Archives, Dr. Quaas and Mr. Klaus-Peter Mertha of the German Historical Museum (formerly Arsenal), Dr. Barbara Grotkamp-Schepers of the German Blade Museum in Solingen, Dr. Martin Westphal of the Rendsburg Historical Museum, Frigate Captain (ret.) Helmut Schmeling and staff of the Naval Officers' Organization, Mr. Eberhard Schmidt of the Military History Training Center of the Mürwik Naval School, Dr. Joachim Niemeyer of the Rastatt Military History Museum, and the staff, especially Mr. Koziura, of Defense Library II in Hannover.

Thanks go out also to Stefan Stefanski for his help in various research.

But the authors want particularly to thank their publisher, Mr. Peter Tamm, for his generous support, which made the appearance of this book possible.

Questions and data about the contents of this book can be sent to the authors—Claus P. Stefanski for the period up to 1918 and Dirk Stefanski for the period since 1919—via the publishers.

Claus P. Stefanski and Dirk Stefanski

Abbreviations

A.K.O.	Allerhöchste Kabinettsorder: All-highest Cabinet Order (from the Kaiser's office)
AMB	Army Museum of Brussels
AVBl	Armee Verordnungsblatt: German Army Regulation Book
DHM	Deutsches Historisches Museum: German Historical Museum (ex-Arsenal)
DKS	Deutsches Klingenmuseum Solingen: German Blade Museum, Solingen
d.R.	der Reserve: of the Reserve Forces
GFM	Generalfeldmarschall: General Field Marshal
HMR	Historisches Museum Rendsburg: Historical Museum of Rendsburg
I.O.D.	Infanterie-Offizierdegen: Infantry officer's sword
kgl.	königlich: royal
K.M.	Kriegsministerium: Ministry of War, Department of Defense
K.M.V.	Kriegsministerielle Verfügung: Available from the Ministry of War
M.A.	Marine-Artillerie: Naval Artillery
MDG	Museum für Deutsche Geschichte: Museum for German History (ex-Arsenal)
MOV	Marineoffiziervereinigung: Naval Officers' Organization
MSM	Marineschule Mürwik: Naval Academy of Mürwik
MVBl	Marine-Verordnungsblatt: German Navy Regulation Book
MWBl	Militär-Wochenblatt: Weekly Military Newspaper
n/M	neues Modell: new model
PS	Privatsammlung: Private Collection
RGBl	Reichsgesetzblatt: Reich Law Journal
R.M.A.	Reichsmarineamt: Reich Naval Office, Department of the Navy
s.-h.	of Schleswig-Holstein
S.M.S.	Seiner Majestät Schiff: His Majesty's Ship (HMS)
SPT:	Sammlung Peter Tamm: Collection of Peter Tamm, Scientific Institute for Maritime and Naval History
STJ	Sammlung Thomas Johnson: Collection of Thomas Johnson
St.O.B.	Station Ostsee, Befehl Nr. ...: Baltic Sea Station, Order number ...
WGM	Wehrgeschichtliches Museum Rastatt: Military Historical Museum of Rastatt
WR	Wilhelmus Rex: King Wilhelm
z.S.	zur See: at sea (Kapitän zur See = Sea Captain)

Explanation of Symbols (for example, p. 14, upper left)
The designs included in some chapters represent the naval emblem of that time.

Notes:
Quotations are printed in italic type.
In the historical sources, the sabers of naval officers are noted as naval sabers as well as naval officers' sabers.

Early History

When one speaks of the beginnings of a navy in Germany, one naturally goes back to the 17th century. At that time the history of the Brandenburg-Prussian Navy will generally begin. As for the exact date, there are varying views. The authors prefer 1657, the year when *Elector Friedrich Wilhelm of Brandenburg and Prussia,* known as *the Great Elector*, equipped several armed vessels.

To build up his fleet, he summoned capable naval men from Holland. One of them, formerly an admiral in Holland and later a privy counselor in the service of the electors of Brandenburg, *Aernoult Gijsels van Lier,* wrote a memorandum in 1658, the *Consilium maritimum,* in which he suggested to the Elector that he be promoted by the Holy Roman Emperor to "Reichsadmiral," or "Admiral General." By combining all the German sea forces under a single flag, a sort of Imperial Navy was to be founded. Two years later he again urged the foundation of an Imperial Admiralty, but the Great Elector did not promote either of these suggestions. After that more than 180 years passed before the idea of an Imperial fleet was taken up again.

Several years later, on March 16, 1675, the Great Elector enlisted another Hollander, *Benjamin Raule,* for his maritime activities. Raule rented him a fleet—of varying size—for various events and missions.

The Great Elector promoted not only nautical, but also colonial activities, for which he needed ships urgently. For example, in 1682 a trade depot was established in Guinea, on the west coast of Africa, followed by others in the course of time. It must be noted that these colonial depots were, to be sure, promoted by the state, but actually established on a private basis by the African Company, a stock company founded for that purpose in which the Elector invested 8000 Taler.

In 1684 the Elector first did without the rented fleet by purchasing eight ships from Raule for 109,340 Taler, though only seven were actually delivered. Koenigsberg, Pillau, and Emden were planned as bases for these ships. But Raule was still available to him, and when the Great Elector died on May 9, 1688, the

1 A lower officer of the Electoral Brandenburg Navy. The model of the sidearm has not yet been determined. C.S. Archives

Brandenburg-Prussian Fleet, including Raule's ships and those of the Brandenburg African Company, numbered 32 ships with some 300 cannons.

Uniform dress and standardized armament in the present-day sense did not exist for that navy at that time. Still, it is noted in a regimental history that members of an infantry regiment (1658-59) were to provide an outfit for the fleet. They are said to have worn a kind of "blue uniform"—J. Zienert[1] calls it "clothing of rank"—and a sword. Picture 1 shows a figure which offers an impression of a seafarer from the time of the Great Elector.

The son of the Great Elector, *Friedrich I,* who named himself King in Prussia, continued to have a fleet, but because of financial shortages he decreased the number of ships regularly, so that toward the end of 1700 the navy consisted of only five seagoing ships, and finally, when the king died on February 25, 1713, there was no longer a single serviceable ship at hand. No data on uniforms or weapons have been located.

His successor, *Friedrich Wilhelm I,* known as the *Soldier King,* broke with his ancestors' naval policy as soon as he came to power. He characterized their maritime interests scornfully as "delusions." As a result, he did not support a navy.

And *Friedrich II, The Great,* who took over the regency on May 31, 1740, declined on principle to establish a navy. As he saw it, neither the financial means that the state could produce nor the numbers of men available to staff a navy in addition to an army were available. This is all the more astonishing since, under his rule, extensive coastal lands—namely East Friesland and West Prussia—were acquired by Prussia and would have justified the existence of a navy. To be sure, in 1757 Friedrich had a coastal flotilla—in present-day terms, a coast guard—established to ward off Swedish ships. It has become known by the name of the Stettin Flotilla. Official acts from this time are no longer in existence and thus cannot give any information about the uniforming and arming of the crews, but in the Berlinische Nachrichten on April 14, 1759,

2 At right is Royal Prussian Ship Captain Nettelbeck. —C.S. Archives

it was reported that *the commanders of the fleet, who held the rank of Captain,* wore *completely blue officers' uniforms, hats with broad gold braid, as well as field symbols* (that is, the portepee), these being the symbols of officers or higher officials in those times. Since a portepee was worn by the commanders of ships and boats, they must have worn a sword or saber as well. We can only speculate as to the type of weapon, but this will not be done here.

The Stettin Flotilla was beaten on September 10, 1759, by the Swedish Army's fleet, which was, indicentally, not the same thing as the Swedish Navy. Three Prussian ships were able to escape. As of February 1760 they were included in the newly formed second Stettin Flotilla. The captains of the new flotilla likewise wore uniforms and field insignia. Very little was reported about the flotilla, but it is noteworthy that in 1761 it carried out a successful mission and captured two ships—one of them the *Mars*[2]. In its best days the flotilla attained a strength of two frigates, three galleys, two bombardment barges, five small "Espingen," and two longboats, with a total crew of 504 men.

In this respect it should not be ignored that *Joachim Nettelbeck,* who became famous later as the citizen adjutant and defender of Kolberg, received a certificate as Royal Prussian Ship Captain in 1769. In his biography he explains that this was the first such rank in Prussia. He also notes that he was also given permission to wear the royal uniform and a saber with a portepee.[3] At the age of seventy, on December 25, 1808, Nettelbeck was honored by the king with the permission to wear the uniform of an admiral, in honor of his achievements in the defense of Kolberg during the Wars of Liberation, on the recommendation of the renowned Prussian *General Neidhart von Gneisenau.*[4] Picture 2 affords—subject to questions about the accuracy of illustrations in those days—an impression of his uniform and saber (from left to right: General von Gneisenau, von Schill, and Nettelbeck).

Friedrich Wilhelm II (1786-1797), the nephew and successor of Frederick the Great, developed no particular maritime interests, so that this period of time can be ignored here.

To sum things up at this point, it is definite that at the beginning of the Brandenburg-Prussian naval history a fleet was established whose uniforms and arms, which were customary for both officers and men at that time, were neither centrally ordered by the state nor regulated by instructions. On the other hand, this does not rule out the possibility that a ship's crew, on their comnader's initiative, were not uniformly dressed and equipped. This state of affairs remained into the nineteenth century.

The First Prussian Navy (1806-1847)

Friedrich Wilhelm III (1797-1840) also declined at first to promote a navy. Only after Napoleon had conquered Germany was he compelled to take another attitude toward a navy.

When Napoleon had beaten the Prussian army at Jena and Auerstedt in 1806, the king had to flee to East Prussia. As a result of this event, the Prussian state treasury also had to be evacuated from Berlin. A water route was chosen for this transport mission, and command was turned over to the pilot commander of Pillau. At the same time, he was issued a "sea uniform." It consisted of *a blue coat with gold epaulettes and two rows of gilded anchor buttons, white cest, long white trousers, the usual civilian portepee and a three-cornered hat.*[5] Unfortunately, nothing was reported as to a side-arm, which undoubtedly was worn.

At that time the position of Governor-General of Prussia was turned over to the Prussian Lieutenant General *Ernst Friedrich Wilhelm von Rüchel.* He had gained attention through his memorandum on Prussia's coastal defenses, which resulted from a trip along the Pomeranian and Prussian coasts in 1796. In his function as governor-general he authorized, on April 3, 1807, the equipping of a "small royal flotilla" to protect the Frische Haff in East Prussia. Five patrol ships, several provision ships, and nine other ships were built. At his request, the captains of all these ships were allowed by the king, via an order of April 20, 1807, to have a uniform. *A blue coat with blue lining, red trim, cutaway lap, with a row of yellow buttons, on which there are anchors, down the front, further with gold epaulettes, likewise with an anchor, plus a white vest, blue trousers, a saber with black shoulder belt, on which in the front on the breast a yellow shield, likewise with an anchor, is located.* Picture 3 gives an impression of how the uniform and saber must have looked.

Another "royal cutter flotilla" was established to protect the city of Danzig. On March 29 its leader received the "sea uniform" already described above. This flotilla was disbanded in May 1807, as it was no longer needed after the conquest of Danzig by the French.

In 1812 Napoleon ordered the equipping of so-called toll ships, which were armed with guns. These ships and their captains, though, were not subordinated to the military authorities, because their official tasks were overseeing the newly-passed trade and financial laws (the Continental System opposed to Britain), protecting the

3 The captain of a Prussian patrol ship wears a saber, the model of which can no longer be determined, though it may be a fusilier officer's saber. — Koehler-Mittler Publishing Group

permitted trade, and preventing smuggling. On the other hand, the captains, in their positions, were granted "all rights and privileges of a naval war commander," which is why they are noted here. On May 2, 1812, the patrol-ship captains were assigned the following uniform by State Chancellor *von Hardenberg: A water-blue coat with white collar, facings, cuffs, trim and underclothing (vest) with gilded buttons, gold epaulettes with the monogram of His Majesty the King, gold and light blue portepees and cordon, plus a saber in a bandoleer, on which was located the attribute of the action and voyage, namely a staff of Mercury plus an anchor, likewise with the monogram of the King, in a medallion.* On May 30, 1814, the Peace of Paris ended the war. The Prussian ships were then either used to guard harbors, docked somewhere, or sold. In the last war against Napoleon in 1815, no new ships were put into service.

In 1815, on the basis of compensation promised in the peace negotiations, Prussia gained the formerly Swedish Pomerania, along with a small nucleus for a new navy. One armed schooner and six cannon boats were gained, and two Swedish officers volunteered to join the Prussian army. They were Captain *Longé* and First Lieutenant *Murck.* In the rank and quarter lists of 1817 they, for a long time the only naval officers in the Prussian army, were listed as members of the Government and Commandery of Stralsund, meaning as members of the army and not as an individual fighting force. The uniforms of the naval officers, determined by an order of July 12, 1816, also had an army look, for the order stated: *I hereby declare that naval officers taken into my service are to wear the uniform cut like the uniform of infantry officers, with a dark blue cloth collar, bearing embroidery like the enclosed model and a red piping, with red Brandenburg cuffs without embroidery, red shoulder pieces and blue lining, and leave it to them to make such known.*[6] Nothing was said of a sidearm. But on the basis of uniform comparison with those of the infantry officers on the one hand, and a picture by P. Pietsch (Picture 4) on the other, the Prussian infantry officer's sword could have been worn, unless the officers wore uniforms on hand from previous service. The idea that a special sidearm would have been introduced for just two soldiers can be ruled out.[7] In other ways too, the two officers were treated like members of the army. Longé gained the ranks of army officers up to that of colonel (1843). Only in 1837 was a small change made for the officers; from then on they were listed in a special section of the rank and quarter list as "naval officers" in Stralsund.

With the short-term establishment of the coastal flotillas, as already described, uniforms as well as a type of weapon, this being a sword or saber, for their captains were determined for the first time in Prussia. The form of the uniform essentially was based on that of the army, even though it included maritime attributes, such as anchor buttons. The same applies to their equipping with sidearms. Available weapons were used, since neither the necessary time to introduce special models nor the interest in them existed. It is also imaginable that the creation of a special model was avoided in order to avoid unnecessary expenses, and the officer in question was allowed to choose for himself; for officers as a rule were expected to clothe, equip, and arm themselves.

4 A naval officer aboard the schooner Stralsund. His sidearm is very probably the infantry officer's saber of the Prussian army. —C.S. Archives

The New Prussian Navy (1848-1866)

In 1840, *Friedrich Wilhelm IV* (-1860) became King of Prussia. He took up anew the idea of a navy, which also ought to be able to operate on the high seas.[8] At first he ordered the building of a school ship, the corvette *Amazone*, to train helmsmen. This ship—in an unusual hybrid role—was to serve as a representative of Prussian shipping and be equipped as a warship, but otherwise as a school ship for the training of civilian as well as military nautical personnel. It was subordinate to the Ministry of Trade and Commerce. The ship, armed with cannons, was to sail under the Prussian flag, though.[9] Along with its military role, the training was chiefly civilian. The command of the ship was at first given to a Danish officer, the sea captain von *Dirking-Holmfeld,* who was replaced in 1846 by the Dutch Kapitänleutnant *Jan Schroeder*. By order of January 4, 1844, the crew of the *Amazone* were given a specified uniform. Their arms were to be a daggerlike deer knife with portepee for the officers (see Picture 5), *for the cadets a dagger, for the training master a bayonet, for the chief fireworks or boatswain a bayonet or a dagger with portepee.*[10] In 1848 the *Amazon* was assigned to the Ministry of War as a school ship within the framework of Prussian naval reorganization. At the same time, three naval officers were taken on.

5 Officers of the corvette Amazon with deer knives. —C.S. Archives

At the beginning of 1848 Denmark possessed German Holstein. Thus on April 4, 1848, the German Federation declared war on Denmark—which was also a member of the Federation. In a countermeasure, German ships, especially a number of German merchant ships, were brought to Copenhagen by Danish ships. Later the German coast, and the harbors in particular, were blockaded. This event caused the question of an effective navy to take on a high value not only in Prussia, but in all of Germany.

Friedrich Wilhelm IV therefore established a Naval Commission on April 17, 1848, and assigned its chair to his cousin, Prince *Adalbert of Prussia,* giving him full authority. The commission was to advise on all naval matters and make appropriate suggestions to the king. As a result, 18 cannon boats and two iron cannon jollyboats were built for coastal protection. The first shallop was launched on August 10, 1848. On September 30 the Ministry of War therefore requested Adalbert, as the responsible authority for naval reorganization, to draw up uniform regulations for the crews of the boats. The regulations which were turned over to the Ministry of War by Adalbert on October 2, 1848, after being accepted orally by the king, can thus be regarded primarily as meant for the personnel of

6 Monument to Admiral Prince Adalbert of Prussia in Wilhelmshaven. —C.S. Archives

these first craft (the cannon-boat flotilla). As for the question of armament, the following was stipulated:

2. For the commanders of the cannon boats.
   Examined Shipmen
   d. Saber

And further:

4. Underofficers 1st Class
   g. Saber on black belt (the model will follow later)
5. *The* **helmsmen** *assigned to this year's training*
   d. The saber of Underofficers 1st Class
6. *The* **captains** *of the merchant marine assigned to this year's training*
   The named merchant captains may wear the saber specified for naval officers.
   The models of sabers will follow later.

The king had also ordered that the naval officers of all ranks could wear an undress uniform. For armaments he ordered: *In addition, they may wear a light saber or, instead of this, a dagger, as both are customary in other navies.—The models will follow.*

On the basis of the cited order it is undoubtedly clear that at the beginning of the Prussian naval reorganization no sidearm models had been determined! (Model) designations in use today, with the year given as 1848, must therefore be abandoned.

Which weapon was meant by the *light saber*? Jan Pieter Puype believes that it is very possible, even probable, that the curved daggers popular at that time were meant as "light sabers."[11] This weapon was actually a "saber made smaller," with an overall length of about 39 cm, which did not exceed the length of the dagger model introduced in Prussia in 1849.[12] In terms of type, these curved daggers belong neither to the daggers nor the sabers. Since it became customary to include these weapons in the group of naval sabers, which is quite acceptable historically, we shall not discuss the matter at this point.

Likewise, it cannot be meant that at this early date there were actually sabers—instead of curved daggers—worn by the first Prussian naval officers, who had to be recruited for the most part from the civilian merchant marine, as well as the state sea trade organization. It can be assumed from this that these officers did not wear regulation weapons, but rather individually made ones, perhaps made after the patterns of other navies. So it cannot be ruled out that at first even Mameluke-style sabers "turned up" among the Prussian navy. Since the first officers' weapons, as noted, were not originally subjected to any norm, we shall not go into the subject further.

Introduction of a Saber for Naval Officers

During November 1848, Prince Adalbert visited British shipyards and naval bases.[13] The experience he gained there enriched the subsequent considerations of the Prussian Naval Commission and were reflected in, among others, the uniform stipulations, as will be shown below.

On April 4, 1849, instructions were made known—at first in a naval order—about the uniforming and arming of naval officers.

So that the instructions would be known in the entire army, they were published by the Minister of War in the military weekly journal on April 27, 1849, as:

Instructions regarding the clothing and insignia of rank for the navy.

In the instructions it was stated:

4. The Saber

The saber has the English model, but the grip is ivory (bone) instead of fish skin. It is worn as a drawn saber. The saber tassel[14] *is worn on a strap, not on a band.*

Models of the newly-made uniforms and arms, including a dagger and also a saber with a portepee, were turned over on April 18, 1849, by the Naval High Command in Berlin to Commodore Schroeder, who directed the naval units at Swinemünde at the time.

W1 The handle of a British naval officer's saber, easily recognizable by the handle covered with fish skin. (PS)

7 A Prussian naval saber, worn by Prince Wilhelm of Hessen-Phillipsthal (admiral's rank). —MSM Archives

Thus the "Prussian Naval Saber," to use the not quite correct but most often used designation for the naval officers' saber of 1849, was determined in its basic model form, even though it differed in small details from the British pattern.

Here it should be noted that the model weapon, like other articles of clothing and armament, may well have been obtained directly from Britain.[15] The hilt of the model was already designed so that the grip was made of ivory instead of the fish skin customarily used in Britain.[16]

Description: If one looks for concrete data in old orders or old literature on Prussian naval sabers, especially as to their measurements, one will be very disappointed. Thus the founder of the Society for Military Knowledge and eyewitness, Adalbert Mila,[17] wrote, not in great detail, about the sidearms of naval officers:

Naval saber in black leather scabbard decorated with gold fittings, brass bow, ivory grip on black leather hanging bandoleer; on it the silver naval portepee. Thus one can depend only on actual specimens and old photographs still available today.

As is shown by the order, the Prussian naval saber was based on the saber model that was already worn in the British navy as "pattern 1827."

Blade: Like the model, the Prussian saber regularly has a stiff-backed blade. The location of the blade, though, is different. While British sabers usually have a brace, the Prussian weapons have no brace in the lower part of the blade, but have a wedge-shaped back edge. The blade is usually decorated with floral and maritime patterns. In the middle of the decorations there is an unclear anchor. The blade length of the examined weapons varies between 710 and 779 millimeters.

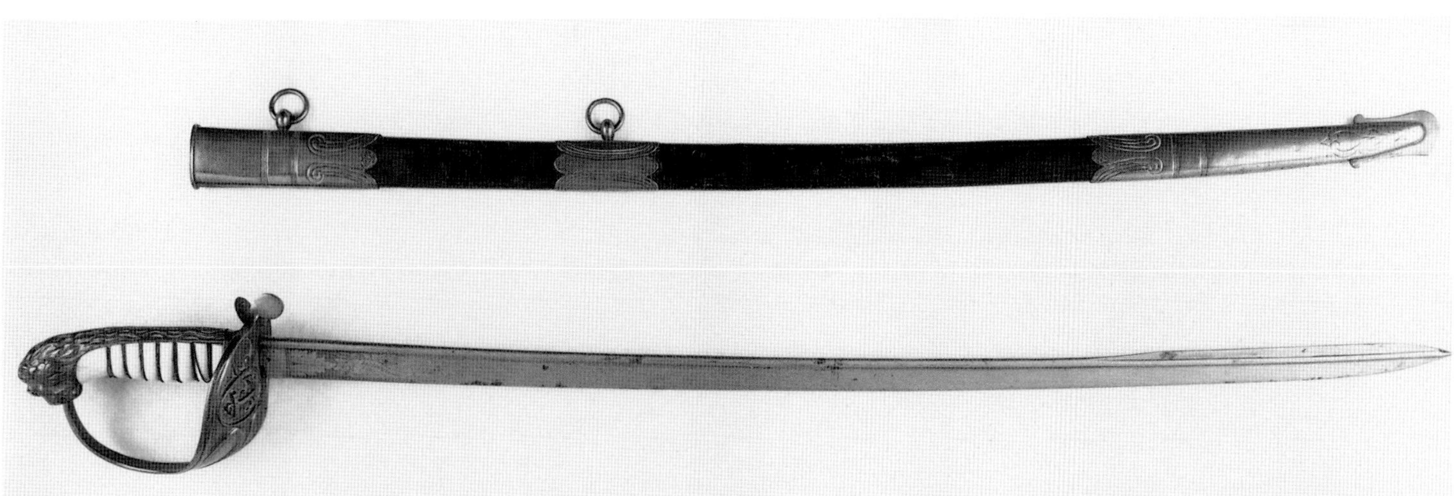

W2 Naval saber model of 1849 for Prussian and later North German naval officers; the stiff-backed blade has a brace, but this is not typical of this model. (AMB)

Left: W 13 A section of the grooved blade of the naval saber shown in W 4 (on page 113), with a free-standing unclear anchor, the typical blade etching of Prussian and North German times. (PS)

Above: W 15 A section of the blade of the naval saber in Picture W 12 (on page 25) with trophies and sailing ship. (PS)

Right: W 20 A typical saber point (blade tip) for Prussian naval sabers, with a ground triangular back. (PS)

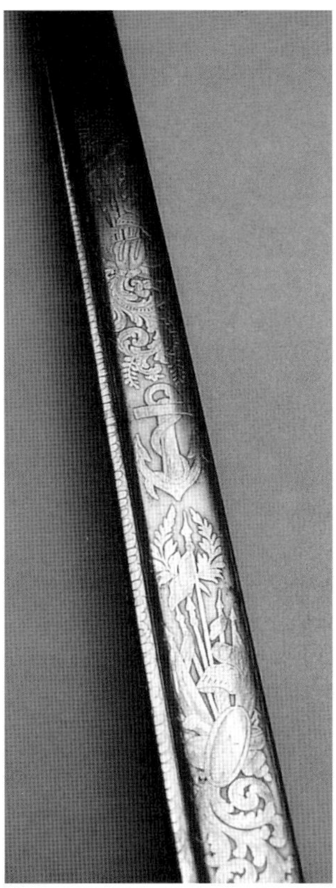

W 14 A section of the naval saber in Picture W 12 (on page 25) with a different type of free-standing unclear anchor, and military trophies below. (PS)

To be sure, in rare cases there are variant blades on Prussian naval sabers, such as stiff-backed blades with a brace or round-backed blades with a groove on either side. There are also blades without decorations. Further, blades still exist today with Damascene steel markings or genuine Damascus steel (see p. 164).

Hilt: The brass hilt[18] consists of a half-basket with fine line decor, with an unclear anchor in a medallion in the center. The Prussian saber differs from the British model, or even its successor in Imperial times in having no crown over the anchor. The snakelike decoration (seen from the top) to the right of the medallion, unlike the lancet-shaped decorative lines to the left, is basically flat, without a bulge. This is a typical feature of Prussian sabers.

The back of the half-basket is fitted with a clasp and thus can be folded back. There are weapons that have a hole in the flap which is made to take a soldered thorn provided on the mouth of the scabbard. This setup serves to hold the weapon and scabbard together. The number of weapons with and without engraved names is about equal.

The form of the bow varies between a semicircle and a rounded rectangle.

W 3 Hilt of a Prussian naval saber, its lion head (after the English pattern) in noticeably large form. (PS)

W 5 Hilt of a Prussian naval saber, the lion's head in a noteworthy form, looking more like a bear than a lion. (PS)

Below Left: W 7 The basket of the saber in picture W 5. The two holes in the guard are customary in the British model; the decoration to the right of the medallion (seen from the top) is flat, typical of the 1849 naval saber. (PS)

Below Right: W 8 The basket of the saber in picture W 11 (p. 113). This saber also has two holes in the guard, while the medallion background is noticeably decorated. The upper part of the bow has a slit, on which the portepee could be hung (in British fashion). In Prussia, though, the portepee was attached in a different way. (PS)

W 9 Basket of the saber in picture W 6 in the color section (p. 114). There are no more holes, and five circles (also called "five balls" in Britain) decorate the anchor. (PS)

In the upper section of the bow, early weapons usually had a slit, and/or two holes in a row on the guard, before the blade. These openings were intended, as in the British weapons, for attaching the portepee. In Prussia, though, it was not customary to use these means. According to observations to date, these features can be ascribed mainly to the older weapons. Over time they lost first the slit and later the boreholes. To be sure, newer weapons occasionally have one or the other feature.

The cap of the grip is adorned with a lion's head. The lion's mane covers the whole cap down to the grip ring. The shapes of the lion-head vary in all the swords known to the authors. Head designs exist with a very high mane and extremely small mouth, and also heads with very marked eyebrows, and sub-variants of this type exist, such as with and without lower jaws.

The grip ring is decorated with channelings. Here too there are "countless" variations.

The grip itself is made of ivory or bone (bleached bone). Seven or eight grooves take up the wire wrapping. This usually consists of three brass cords. In Britain, as already noted, the wooden saber grip was covered with fish skin (pearl shark). Polished ivory grips, as used in Prussia, were used there only by flag officers.[19]

W 10 Basket of the saber in picture W 4 in the color section (p. 113), The anchor has not only the five balls, but also fine decorative lines on the shaft, arms, and flukes. (PS)

W 21 Three upper scabbard fittings of Prussian naval sabers, in varying forms. (PS)

W 22 Four central scabbard fittings of Prussian naval sabers; the shorter types were customary, especially in the earliest days of the Prussian navy. (PS)

W 23 *End fittings of three Prussian naval sabers. The sliders still turn into flanges in the older types. (PS)*

Scabbard: The scabbard is made of leather and has three brass fittings. The mouth and middle pieces always have ring eyelets soldered on.

The vertical decorative grooves of the middle fitting regularly extend over its entire length. It should also be kept in mind that, even if it is only an exceptional case, an undoubtedly Prussian scabbard exists with vertical grooves divided into two sections by horizontal grooves. This actually became customary only in Imperial German times.

The end fitting is asymmetrical and, in early cases, ends in two "cheeks." Newer swords lack these cheeks. The age of the scabbard can still be estimated by the distance of the horizontal decorative grooves from each other at the level of the ring eyelet on the mouth fitting and the tip piece. On older scabbards the grooves are relatively close together, and on the mouth fitting this means right at the level of the upper and lower ends of the eyelet.

Manufacturers and Dealers: Obviously it became popular in the middle of the 19th century for manufacturers not to sign their work anymore. Thus the manufacturer of a weapon is often no longer identifiable today. Only some of the examined specimens can be traced to their makers. In most of these cases a [Wolfsangel] in a crowned coat of arms was found on the dull edge. This symbol was used since the fifties of the last century by the Solingen firm of *H. W. Lang*, which

Technical Data of Examined Prussian Naval Swords

Weapon	1	2	3	4	5	6
Length sheathed	978 mm	918 mm	928 mm	899 mm	875 mm	870 mm
drawn	854 mm	877 mm	858 mm	849 mm	853 mm	
Blade type	2 hollow grooves	back w/ brace	back w/ triangle	back w/ triangle	back w/ triangle	back w/ triangle
Blade length		723 mm	747 mm	724 mm	710 mm	718 mm
Blade width		26 mm	23.5 mm	25 mm	29 mm	28.3 mm
Blade thickness		7.5 mm	6.5 mm	7 mm	7 mm	5.7 mm
Hilt height	130 mm	131 mm	130 mm	130 mm	135 mm	135 mm
Bow type	squared	rounded	rounded	squared	squared	squared
Slit / holes	none / 2	none	none	yes / 2	yes / 2	yes / 2
Attacher	no	yes	no	yes	no	yes
Grip grooves	7	7	8	8	7	7

(Sheathed: the blade is in the scabbard.)

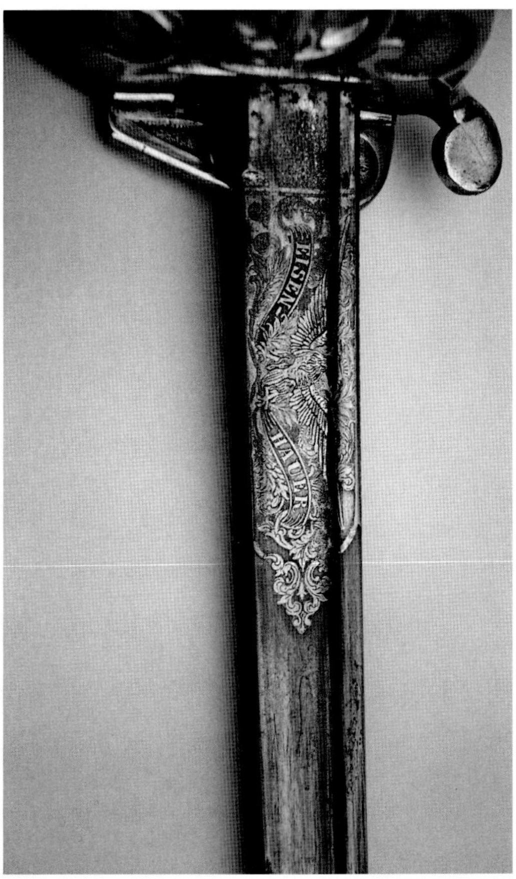

Above: W 12 Early naval saber with manufacturer's symbol, "squirrel in oval." This "trade mark" was assigned to the firm of Friedrich Eickhorn as of 1847 (from Kronau, History of the Solingen Blade Industry). (PS)

Left: W 16 The cutout in the blade showed the Prussian eagle below a crown and the ironworkers' symbol of quality for the blade; this type of blade decoration is rarely found on Prussian naval sabers. (PS)

W 17 Upper scabbard fitting with the mark of the royal supplier W. Lucas of Berlin. (PS)

W 17 Upper scabbard fitting with the mark of the royal supplier W. Lucas of Berlin. (PS)

existed until about 1870.[20] The symbol is also found on the weapons produced by the firm of *Clemen & Jung*, which was founded in 1860.[21] When the symbol was turned over to the firm of Clemen & Jung is not known exactly; Cronau still listed it for H. W. Lang in 1885. Further manufacturers identifiable by the blades are the firms of *Schnitzler & Kirschbaum* (stamped S & K) and *Eickhorn* of Solingen.

Large-scale dealers also immortalized themselves on naval sabers that they distributed. The firms of *W. Lucas* and *J. & E. Eisner,*

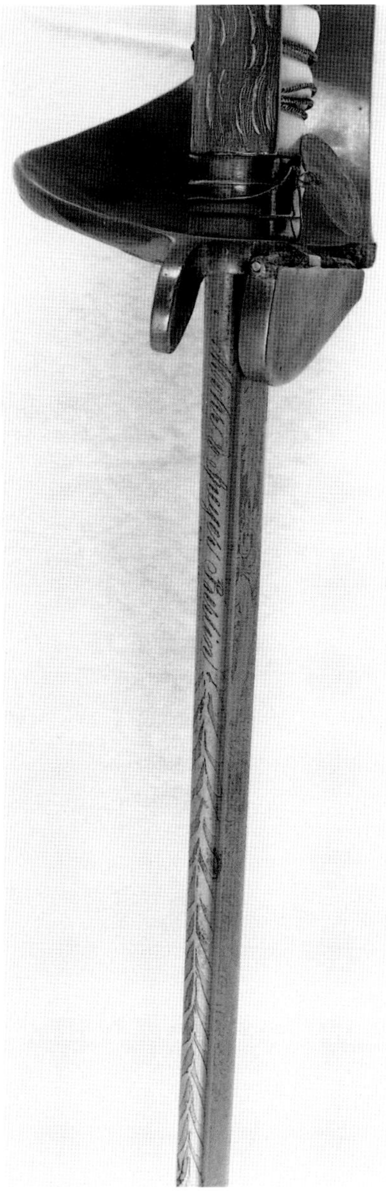

W 18 A dealer's signature on the back of the saber blade (picture W 2, p. 19): Mohr & Speyer, Berlin. (AMB)

W 19 A dealer's signature on the back of the blade (picture W 16, p. 25): Adolf Wiener, Berlin. (PS)

both of Berlin, can be named. In a *Price List of Military Effects and Uniform Clothing of the Court Suppliers and Manufacturers Mohr & Speyer* of Berlin for the year 1857, the naval saber is offered as follows:

> ***1 Saber** etched and fire-gilded hilt, English braced blade. 16 to 18 Thaler.*

Prussian naval sabers, though, were bought and sold not only in German lands, but also in Britain. Thus wrote *von Holleben: Plymouth was our second, if not our first, source of equipment. Unfamiliar with the language, we used the price lists to deal for everything our money could afford, from a paper collar to the most expensive naval telescope, boots and hairbrushes, woolen and linen objects, sweets and cheese, fabrics and **sabers**—everything was better in Britain than in our harbors at home....*[22] In addition, H. W. Lang and Clemen & Jung both delivered their products to Britain.[23] Thus it is not impossible that Prussians bought their sidearms—made in Solingen—in Britain.

Who Else Wore the Naval Saber

In addition to naval officers, there were other groups of persons who were given the right to wear the naval officer's saber. To the officers' group there belonged the auxiliary officers,[24] these being captains, helmsmen, etc., called into service in the navy from

8 Prussian naval saber worn by Captain-Helmsman Sachse of the sea defenses (auxiliary officer). MSM Archives

the merchant marine. For them it was stated in the regulations of April 1849: *Likewise the auxiliary officers wear no saber for the time being, but only a dagger....*

The auxiliary officers seemed not to have been happy about these regulations, for at that time a saber or sword was "the symbol of rank" of the officer's exalted position. After an inquiry by Commodore *Schroeder,* the regulation was given further explanation as early as April 18, 1849, in a message from Prince Adalbert, who was responsible as Commander of the Navy:

When there are those among the auxiliary officers who do not object to being issued a saber, then nothing is to prevent them from wearing it, rather for the aforementioned officers to wear only the dagger was cited only with regard to the matter of cost. In this case it is also not necessary for those to acquire a dagger.

9 Prussian naval saber worn by a junior portepee officer, Chief Boatswain Hildebrandt. MSM Archives

The deck officers developed from the Unteroffizier First Class in 1850. They, like the sea officers, had to provide their own uniforms. As for sabers, it was stated in a comment from the Naval High Command on April 22, 1850, that it had not yet been decided by the Ministry of War whether sabers were to be delivered by them or obtained by the individual, who should receive compensation of 10 Reichsmark. It was also still to be decided whether one who had obtained his saber himself—with the compensation—was allowed to keep it after leaving the service. Finally, on May 1, 1850, it was reported via Prince Adalbert that the deck officers could regard the weapons as their own, what with their—not inconsiderable—additional expenditure.

In this regard it is to be reported that at the beginning of 1849 the Prussian Ministry of War had bought a number of saber blades and scabbards and made them available to Prince Adalbert. The missing parts, these being the hilts, were to be obtained gradually, and the complete weapons were to be turned over to deck officers in return for the costs.

On April 22, 1850, the Naval High Command in Berlin also issued to the Naval Command in Stettin 21 sabers, two of which had white handles, and they added that the handles of the others should be replaced by ones made of ivory or bone. These instructions resulted in confusion at the Naval Command in Stettin. For that reason an inquiry was made on April 25, 1850, as to whether all the sabers were to be fitted with white handles. It was noted that in the summer of 1849 a saber with a black handle was received for the gunner, and another with a white handle for the boatswain, of the *Amazon*. In an order of May 1, 1850, Prince Adalbert therefore declared:

> *Even though I would have liked to see that the sabers of the deck officers had the same handle of ivory as those of the officers, yet I want, in view of the resulting high costs, that the handle, instead of being made of ivory, be black, likewise with yellow flutings, but that in forming the shape of the basket, this, like that of the handle, shall be exactly the same as that of the officers' sword.... The saber will be worn as a trailing saber.*

In the course of time, though, the white handles came to be accepted. It may be determined that the deck officers, who had to buy their sabers themselves, bought them that way so they would differ as little as possible from those of the sea officers.

The naval cadets were yet another group. On January 1, 1850, Prince Adalbert, as High Commander of the Navy, informed the Ministry of War that the king had, among other things, approved the uniform suggestions for the naval cadets. For the first-class cadets[25] the same weapon—thus the saber—was prescribed as for the sea officers. The aforementioned regulation was published with the Ministry of War's directive in the military weekly news on March 17, 1850.

The second-class naval cadets[26] who had earned the "certificate for the level of Unterleutnant zur See" and served a period of three years at sea, could apply through the Commander of the Fleet Division or the appropriate commander, according to an order from the Naval High Command on November 21, 1868, for permission to wear a saber.[27] In military jargon these officers were later referred to as "saber ensigns."

On April 27, 1849, the uniform prescriptions for doctors and paymasters[28] were issued, and on November 23, 1850, those for subpaymasters. They too were allowed to wear naval swords. The same applied since November 23, 1850, to the administrative personnel, as long as they held at least the rank of deck officers.[29] The model was also prescribed for the commander and higher-ranking pilots of the—military—Jade pilots, according to an order of March 3, 1864.[30]

The naval saber was first allowed for officers of the sea battalions in an order of February 6, 1862.[31] Until then they had worn the sabers of fusilier officers.

The sergeants of the sea battalions were allowed to wear the naval saber by order of the Naval Ministry on February 16, 1865.[32] In the order it was stated:

> *In very humble reply to the gracious letter of the Royal High Command of the first of this month, the Naval Ministry approves that the staff sergeants and the sergeants of the sea battalions may now wear the naval saber model as approved*

10 A Prussian naval saber worn by a naval official, Sub-Paymaster Hintze. MSM Archives

for officers, since there are the same reasons for the introduction of this sidearm for the staff sergeants as for its introduction for the officers, and the introduction of the saber for the sergeants seems desirable because of the uniform armament of the troops. The costs of the first equipping of the hereafter required naval saber are to be covered from the sea battalion's savings from their armament repair funds.

In 1858 the "General Regulations as to the Clothing in the Royal Navy" were issued.[33] These regulations included guidelines for the various types of clothing of the naval officers, sea ensigns, and cadets. The saber was required on principle for dress on board as well as on land. The dagger as an alternative for sea officers was specifically banned for use with formal dress and overcoat on land, that is, the saber was always to be worn then.

11 A group photo of Prussian naval officers.

The Sabers of Sea Battalion and Sea Artillery Officers

The Fusilier Officer's Saber

Toward the end of 1849 Prince Adalbert passed on a spoken statement of the King of Prussia about the uniforming of the Naval Corps, later the Sea Battalion. On January 1, 1850, he shared with the Ministry of War the instructions as approved in principle. For arming the officers and thus also the portepee lower officers it had been ordered: *The weapon of the officer is the fusilier saber of the infantry.* Three months later, on March 17, 1850, the Minister of War published the prescriptions in the military weekly news.[34] The model of the fusilier officer's sword, taken over from the army, had finally been defined there by order of August 7, 1835.[35]

W 24 A Prussian fusilier officer's saber: the early type, as was worn by the officers of the sea battalion until 1862. The shield-shaped carrier hook was used exclusively in the navy, but only until 1881 in the Prussian Army. (PS)

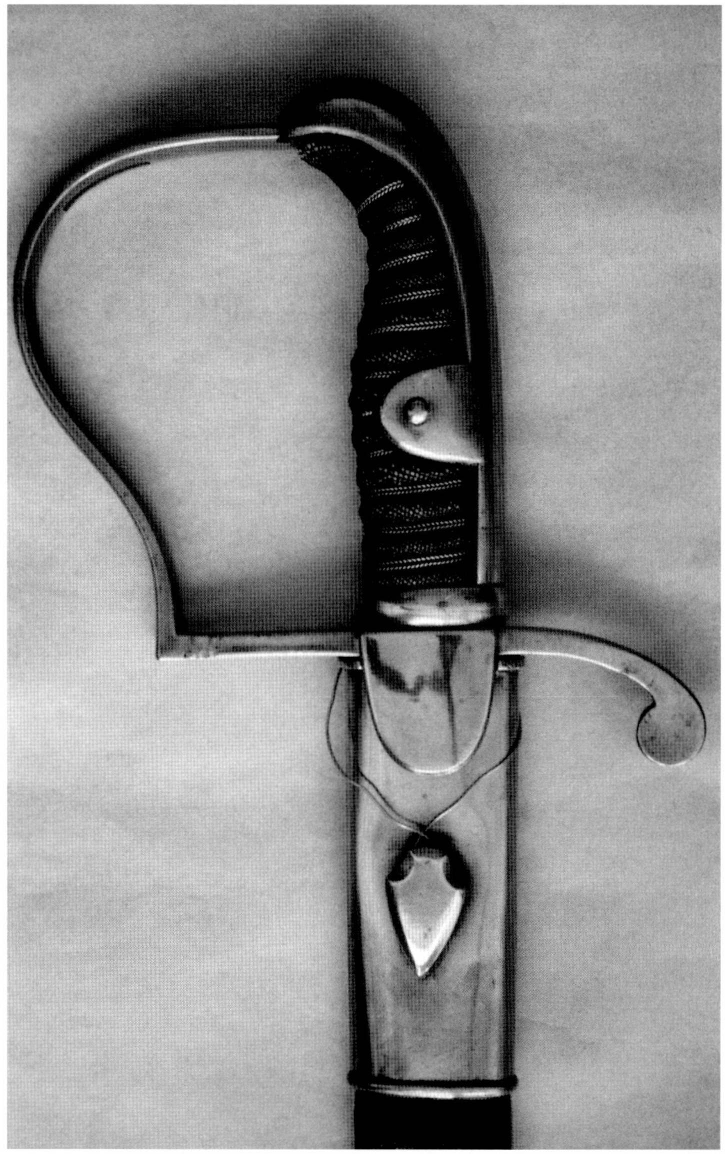

W 24 A Prussian fusilier officer's saber: the early type, as was worn by the officers of the sea battalion until 1862. The shield-shaped carrier hook was used exclusively in the navy, but only until 1881 in the Prussian Army. (PS)

12 A captain of the sea battalion (right) with a fusilier officer's saber. C.S. Archives

W 25 Artillery officer's saber, here as an ordnance weapon for sergeants (army portepee from later times—after 1897). (PS)

For almost twelve years the fusilier officer's saber was worn by the officers of the sea battalion, until finally, by order of February 6, 1862, as noted above, they were also allowed the naval officer's saber. The introduction of the fusilier officer's saber was one of the purposeless decisions that had been made unintentionally in the formation of the new troop unit. Since the saber had to be worn firmly hooked in place there were accidents, especially while walking down gangplanks. The impetus for the armament decision (order of February 6, 1862) is said to have been the fatal fall of *Captain John* aboard the *Barbarossa*.[36]

The fusilier officer's saber then appeared as follows:

Blade: The flat blade, hollow-ground on both sides, is moderately curved, with its point in the line of the back.

Hilt: The hilt is made of gilded brass. The hilt bow, simple and widening to the front, joins the parrying bar at a right angle. The parrying bar has two half-round flanges. The smooth grip cap is held by the smooth grip ring at the bottom; a rivet usually runs from one cap flange to the other.

Scabbard: The scabbard is made of blackened leather with two brass fittings. The upper one (the mouth fitting) has a carrier shaped like a coat of arms, while the lower one (the end fitting) is closed downward by an oval brass button. The more or less long carrying hook often seen on earlier fusilier officers' swords was officially introduced in the army only in 1881 (order of November 14), as the weapon was retired from use by the navy.[37]

The Artillery Officer's Saber

As a result of instructions of October 31, 1857, on November 11, 1857, the "Sea Artillery" was established with a strength of one company and assigned to the Sea Battalion. Its members came from the army artillery. The instructions said of armaments: *The officers of the Sea Artillery retain the artillery saber with the steel scabbard, including a black painted sword-belt with the infantry portepee.* Naturally, this also applied to the sergeants. In 1867 the

Sea Artillery was then removed from its subordinate position to the Sea Battalion. Under the command of a staff officer it became independent, having grown to three companies by this time. On March 1, 1867, the Commander of the Sea Artillery Department requested that the naval saber, so as to present a uniform appearance with the Sea Battalion, now be allowed for the officers of his unit. On March 20, 1867, though, the request was rejected by Prince Adalbert for reasons of cost. In 1877 the Sea Artillery was disbanded and two seamen's artillery units were formed from it.[38] Their uniforms and armaments were "naval" from then on.

The lower officers' weapon with portepee, which was officially specified, will be described below. The officers had to provide their equipment themselves, and could wear a weapon that varied in form and decoration. For that reason the possible variations will not be treated here.

Blade: The curved blade has a hollow neck on each side. The back of the blade is concave to the line of the scabbard (Pandour point).

Hilt: The hilt is made of brass. Of the two angular parrying-bar flanges, the outer one shows two crossed cannon barrels; the inner one is smooth. The grip, consisting of a wooden core, is covered with fish skin and wound with two silver cords. The grip cap is decorated with a lion's head, and is even ornamented with a floral decor on the back and the handle-cap flanges.

Scabbard: The cast steel scabbard has two carrying bands with brass rings.

The German Reich's Fleet (1848-1852)

As has been noted already, on April 13, 1848, the German Bund declared war on Denmark for its occupation of Holstein. Just one day later, a request to form a fleet was made to the Bundestag in Frankfurt. As a result, an office was formed that was fully empowered to handle negotiations for the building and purchasing of warships. The basis of the maritime activity was to be the so-called St. Paul's Church Constitution approved on March 28, 1848. In Paragraph 19 the regulations for a naval fleet were stated. The complicated legal conditions of the time that affected the formation of the fleet do not need to be dealt with here.[39] They did, though, result in a limitation of the navy.

On May 11, 1848, the so-called "fifty office," which was responsible for the tasks of the National Assembly until the election, issued a *call for the founding of a German naval fleet.* Unprecedented enthusiasm for a fleet broke out. All possible groups of society and many personalities placed themselves in the service of this national issue. The National Assembly, which convened on May 18, formed a "naval committee" and approved six million Taler for the founding of a German fleet.

The Reichs' Trade Ministry became responsible for the establishment of the fleet. At its head was the Bremen citizenry's president *Arnold Duckwitz,* who was also the leader of the naval department. Assigned though not subordinated to him for factual advising was a "Technical Naval Commission."

On request from the Reich's administrator, *Archduke Johann,* the King of Prussia released Lieutenant General Prince Adalbert on November 17, 1848, to direct the Technical Naval Commission, which he chaired until the end of February 1849. He had already drawn attention by writing a *memorandum on the formation of a German naval fleet.*[40]

The Commission worked out, among other things, the *instructions regarding the uniforming of the officers and men of the Reich's*

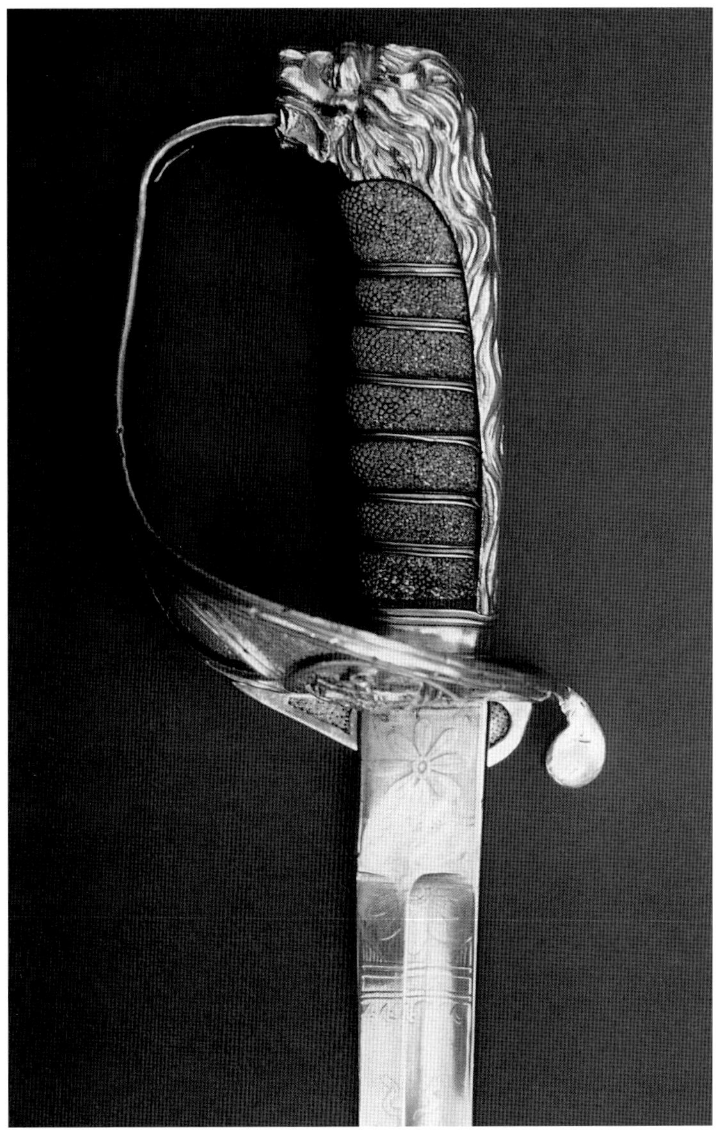

W 26 The naval saber of an officer of the Reich Fleet of 1848-1852. The blade has no reinforced back, but has two hollow channels; the handle is wrapped in rayskin.

W 27 Hilt of a naval officer's saber of the Reich Fleet; the medallion includes the uncrowned double eagle as the symbol of the Reich's power, superimposed on an unclear anchor; the decorative scrollwork to the right of the medallion (top view) is, unlike the Prussian naval saber, not flat. The bow is cut out as was the Prussian weapon. (PS)

Navy on March 1, 1849. In Paragraph 1 it states as to sidearms: *Saber after the British naval pattern, on a black lacquered belt with gilded fittings, on the lock the unclear anchor. Sword tassel in the national colors.*

The officers of the Reich's Fleet also had to provide their sidearms themselves. The same applied to the auxiliary officers. In all, there were 29 officers, 26 ships' ensigns, and 16 genuine sea Junkers.

The medical personnel (nine doctors and one druggist) and the administrative personnel were to wear a uniform similar to that of the sea officers. Since no more specific description of the sidearm had been made in the instructions, it may be assumed that these personnel also wore the same model of saber.

Description: The officer's saber of the Reich Fleet is illustrated in an old specialist magazine and described as follows: *Saber after the British naval model on black lacquered belt with gilded fittings, on the lock the unclear anchor. Saber tassel in black-red-gold colors.*[41]

The Reich Navy's officer's saber was also based on the model that, with slight changes, was used by the British Navy since 1827.

Blade: A saber still extant today has a blade with rounded back and carp-tongue point. It is decorated on both sides with arabesques in neo-Gothic style. In the middle of the decoration is an unclear anchor with a double-headed eagle, over it a radiant wreath and behind it a laurel branch. Set in the dull edge is a brass marking "PROOF" (tested). Another blade has not only arabesque designs but also maritime and military sylbols, such as unclear anchors, cannon barrels, and three-cornered hats. Its dull edge is decorated with flowers.

Hilt: The—gilded[42]—brass hilt is a so-called three-quarter basket, adorned with elegant line patterns, with an unclear anchor with two-headed eagle set in a medallion. The form of the bow is more like a rounded rectangle. In the upper area of the bow is a slit, and one weapon has two holes side by side in the guard—before the blade. The inner side of the basket can be folded.

The grip flap is decorated with a lion ring. The lion's mane covers the entire grip flap to the ring. The grip ring is decorated with flutings and/or raised rings. The grip itself consists of a wooden core covered with fish skin, as was customary in Britain. The fish skin, as old traces of color show, was whitened. Six or seven grooves hold the wrapping wire; this consists of three strands.

Technical Data of examined officers' sabers of the Reich Fleet:

Weapon	1	2
Length sheathed	962 mm	
drawn	925 mm	935 mm
Blade length	790 mm	797 mm
Blade width	28 mm	28 mm
Blade thickness	7.5 mm	7.5 mm
Hilt height	140 mm	138 mm
Bow shape	straight	straight
Slit/holes	yes/two	yes/none
Attacher	no	no
Handle grooves	7	6

Scabbard: The scabbard varies in two ways from the Prussian type. One difference is in the middle fitting, which is not decorated with vertical lines as the Prussian saber was, but with horizontal lines at the level of the ring holder. The fitting thus resembles the weapon of the Imperial era. An important difference is the symmetrical outward curves of the carrier, as are often to be found in Prussian military sabers.[43]

W 28 Tip fitting of the naval officer's saber of the Reich Fleet. (PS)

W 29 Deck officer's saber of the Reich Fleet of 1849, grip cap without tabs. (PS)

For the deck officers, the aforementioned prescriptions for sidearms say simply: *Sword according to model, with a tassel in the national colors.* Now one could presume from this that the Reich fleet's deck officers were supposed to wear the same model as the sea officers. But why, then, was this weapon not referred to in the text? Obviously, the two models must have differed from each other.

It is known that on April 3, 1849, twelve examples were made by the firm of Schnitzler & Kirschbaum (S & K) in Solingen. With the accompanying belt, a weapon cost eight Thaler and 16 silver Groschen then. The sabers were all delivered by May of the same year.

One of these weapons, or the model that was presented to the Naval Technical Commission was, in any case, sent on to the model collection of the "Reich Naval Ministry." When at the beginning of 1867 all remaining so-called "federation furniture" was to be auctioned off publicly as directed by the Federation Liquidation Commission, it was decided to send some items of historical interest to the Germanic National Museum.

W 30 Lion's head of the grip cap of a deck officer's saber (picture W 29). (PS)

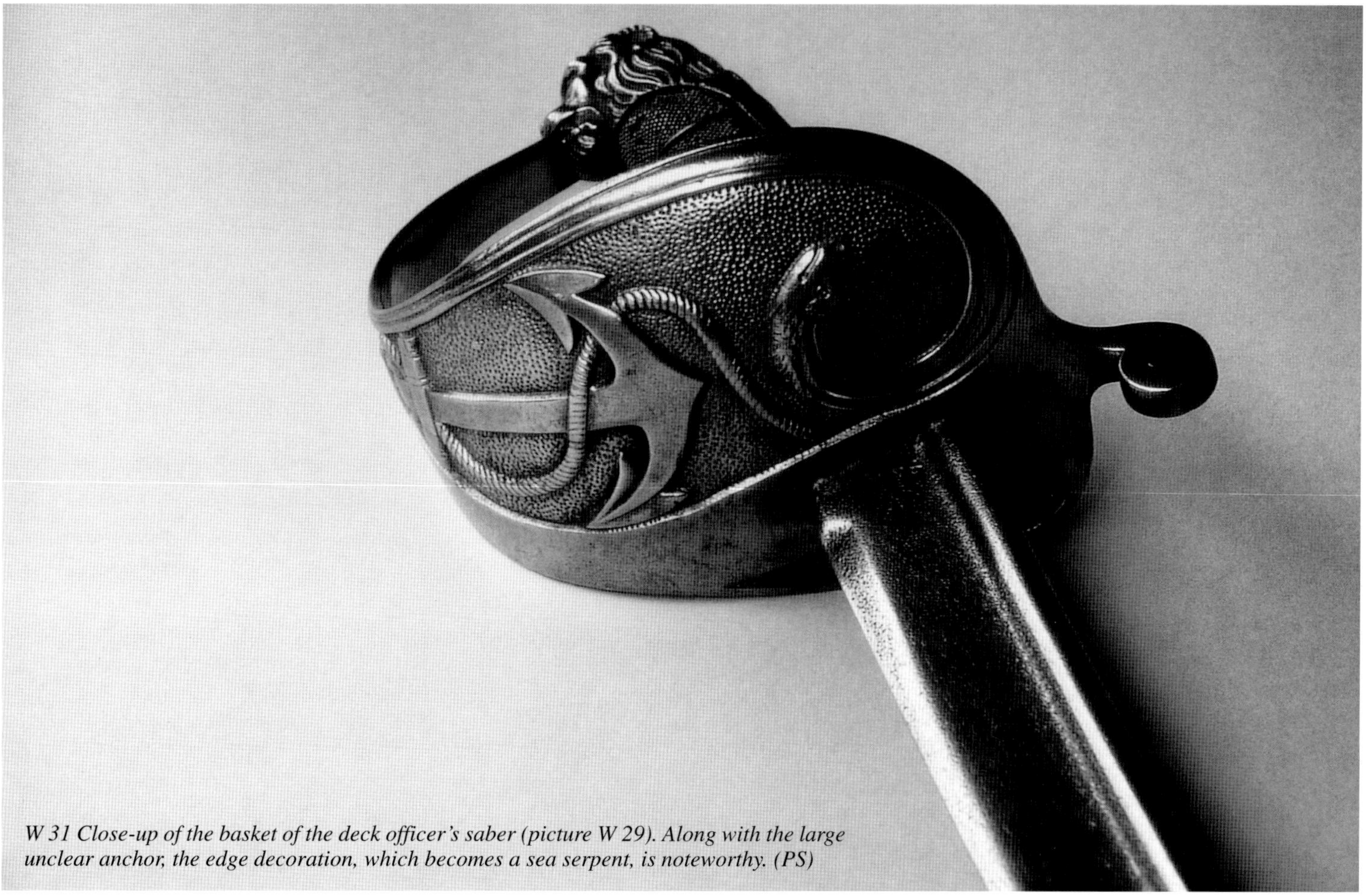

W 31 Close-up of the basket of the deck officer's saber (picture W 29). Along with the large unclear anchor, the edge decoration, which becomes a sea serpent, is noteworthy. (PS)

Technical Data of deck officers' sabers of the Reich Fleet of 1849:		
Weapon:	1	2
Length sheathed	... mm	... mm
Length drawn	880 mm	880 mm
Blade shape with brace	reinforced with brace	reinforced
Blade length	754 mm	
Blade width	30 mm	
Blade thickness	6 mm	
Hilt height	126 mm	
Bow shape	C-shaped	C-shaped
Handle cap	Lion's head	Lion's head
Handle grooves	7	7
Slit/holes	none	none
Manufacturer	S & K	S & K

Along with the firearms, there were also a boarding saber, a boarding pike, and also a model of the deck officer's saber included.[44] Luckily, the saber has been preserved to this day, so that it can be described below.

Description:

Blade: The gently curved blade has a reinforced back with projecting brace.

On the blade of the model are an uncrowned double eagle (outside) and, on the other side, a clear anchor.

Hilt: The brass hilt consists of a closed basket, decorated with a large unclear anchor. The outer rim of the basket is finished with decorative lines which—in the lower part of the basket—end with a sea serpent's head.

The grip cap is decorated with a lion's head.[45] The lion's mane covers the handle cap only in its uppermost part. No grip cap tabs are present.

The inner part of the basket cannot be folded.

The grip ring has been left smooth.

The grip itself consists of a wooden core wrapped in fish skin. Seven grooves take up the wire wrapping. This consists of two entwined brass wires.

Scabbard: A description of the scabbard, unfortunately, cannot be given, for all the known weapons no longer have one.

The Hamburg Flotilla (1848)

The great eagerness for a fleet in 1848 also had its effect in the coastal cities. On May 8, 1848, Hamburg shipowners began to make voluntary contributions to convert a sailing ship, the *Deutschland,* to a frigate. One shipowner provided a second sailing ship. Three steamships were also bought, and their civilian personnel were immediately taken into the planned military service. The officers' positions, insofar as they were not filled by the personnel mentioned above, were filled by captains and helmsmen from the merchant marine. The ships, known collectively as the *Hamburg Flotilla,* originally flew the Hamburg Admiralty flag.[46]

On October 15, 1848, the ships were turned over to the Reich, and they were subsequently improved and equipped for war service. The second sailing ship, though, could not be taken over, as it was fully unsuitable for the foreseen use. The ships formed the core of the German fleet. After being turned over, they flew the Reich's black-red-gold flag.

The crews of the ships were probably uniformed even before the ships were turned over,[47] but how the uniforms and, in particular, the weapons may have looked cannot be determined. To be sure, one naval saber exists,[48] doubtless of Hamburg provenance, which can be ascribed with all rights reserved to one of those naval officers. The weapon much resembles the Prussian naval saber, but differs in having a stylized white-metal Hamburg Gate superimposed on the unclear anchor. Also, the grip of the weapon is not made of ivory, as the Prussian type was, but of wood, covered with bleached fish skin.

The Schleswig-Holstein Flotilla (1848-1852)

Originally independent of the Frankfurt Parliament, Schleswig-Holstein began to set up a small fleet and a naval authority in the summer of 1848. The ships numbered sixteen and were equipped with 41 cannons and about 790 men. The crews were assembled from Schleswig-Holstein naval officers and men, amplified by gun crews from the Hanoverian and Prussian artilleries.[49] The Schleswig-Holstein Flotilla had 63 officers and officials. On January 11, 1851, the fleet was disbanded after the ships had been turned over to Denmark and the crews taken over by Prussia. Their deeds were always given high praise.[50]

13 An officer of the Schleswig-Holstein Navy (center) with saber (drawing by H. Knoetel). WGM Archives, Rastatt.

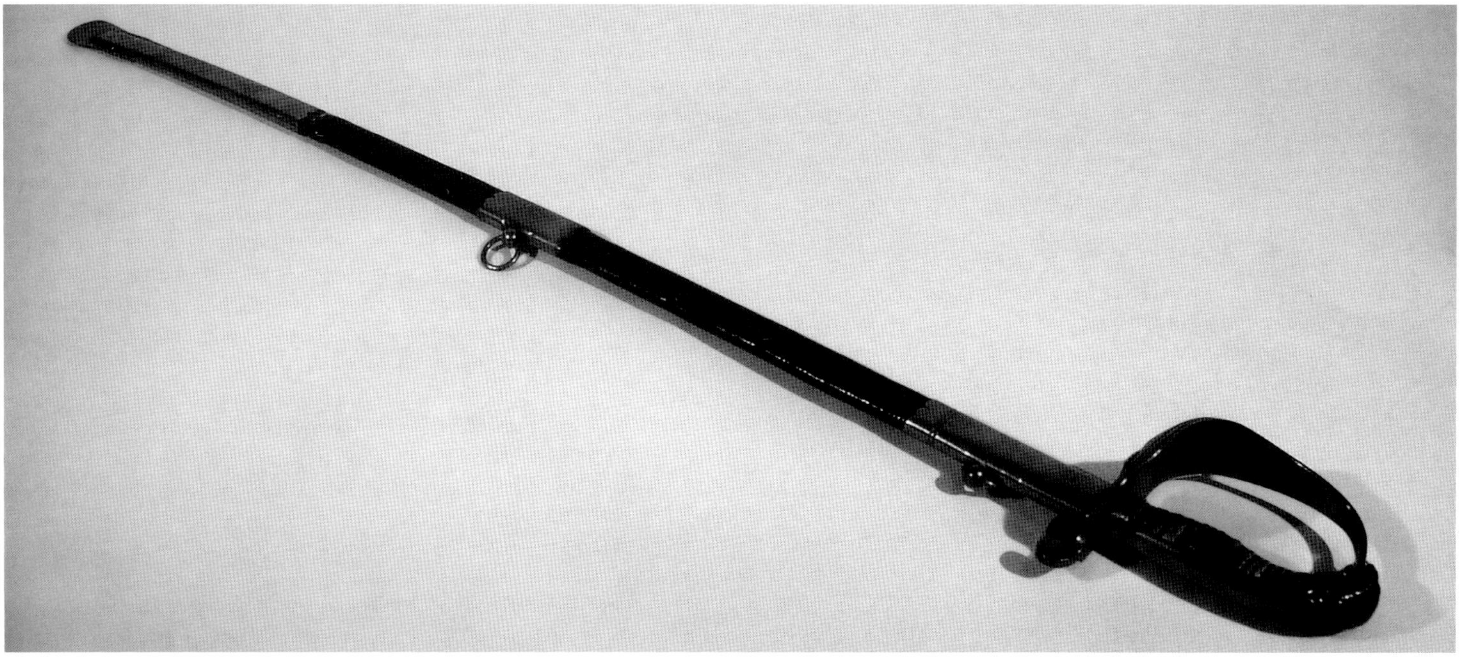

W 33 Naval saber of an officer of the Schleswig-Holstein Navy in 1848. (HMR)

No uniform regulations have been found to date.[51] Rather, it was on the basis of a few uniform pieces and illustrations that Gerd Stolz undertook to reconstruct the uniforming of the Schleswig-Holstein Navy.[52] As to the officers' sidearms, he determines: *Saber on black lacquered leather belt.*[53] Herbert Knoetel, one of the great German uniform scholars, also portrays, on a uniform table, three members of the Schleswig-Holstein Navy. The central figure represents an officer, who wears a yellow-mounted saber in a black leather scabbard with three fittings.

In a private and a public collection there exist two almost identical saber models, one of which can be identified by its first owner's inscribed name as belonging to a Schleswig-Holstein naval officer. In the public collection the same model, which was obtained in 1864, is also attributed to Schleswig-Holstein by its old inventory card. Since no other pattern has to date been identified as being from Schleswig-Holstein, these weapons can be regarded as naval officers' sabers of Schleswig-Holstein. It is noteworthy that the basket is extremely like that of the Reich Fleet deck officer's saber. Why the Schleswig-Holsteiners followed this and not the British model is not clear.

Left: W 34 The hilt of the Schleswig-Holstein saber (picture W 33); the grip cap ends in an eagle's head. The ray skin of the grip was colored white. (HMR)

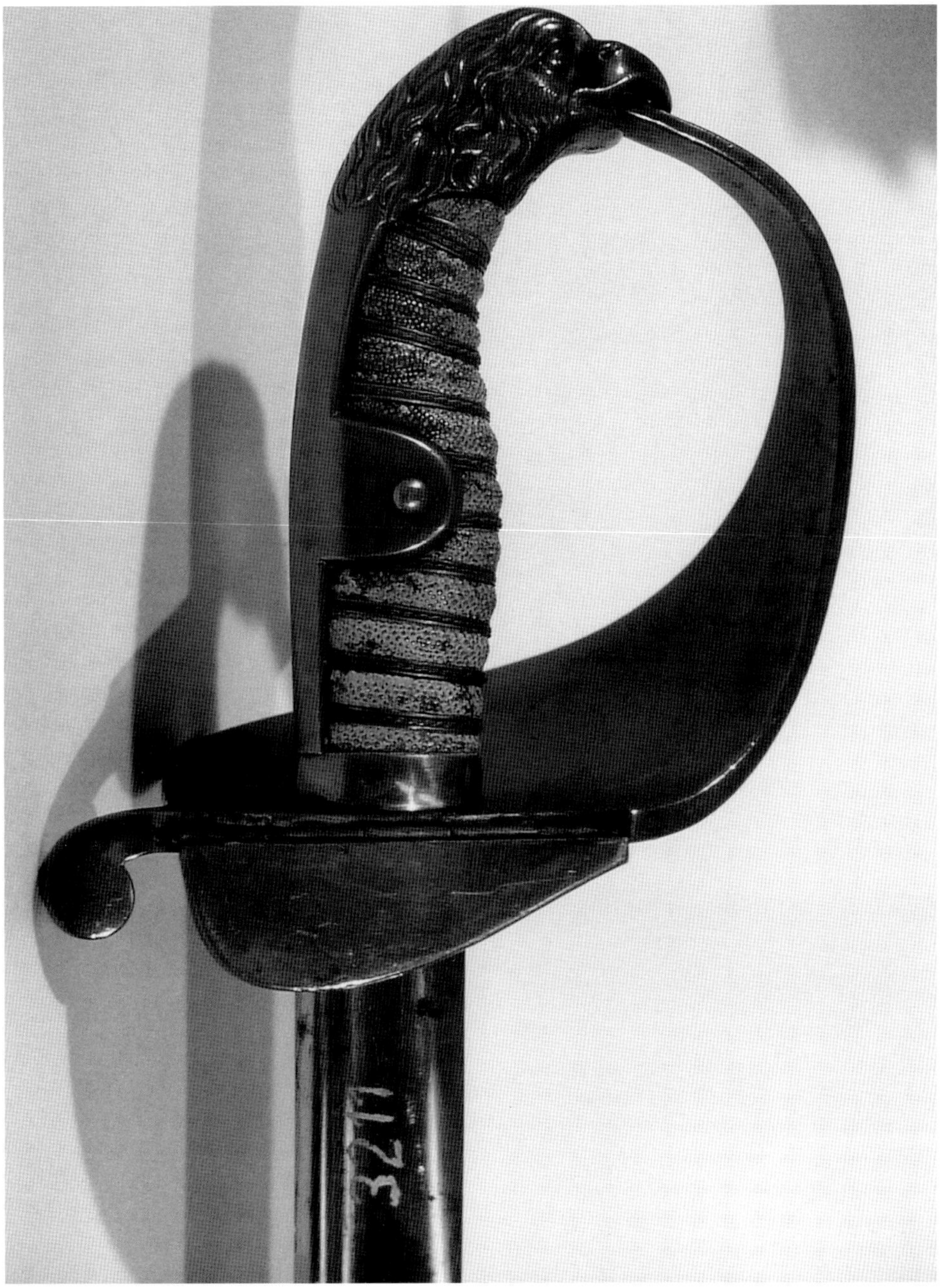

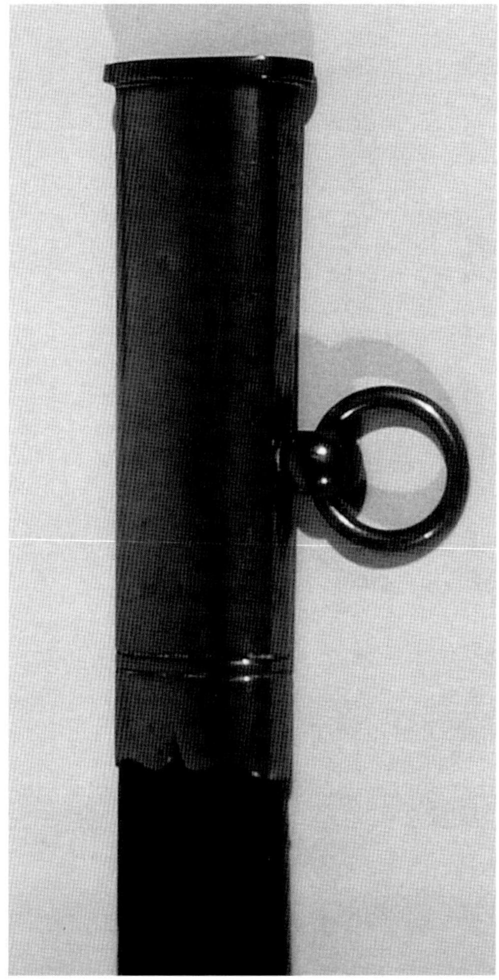

W 37 The upper scabbard fitting of the Schleswig-Holstein naval saber (picture W 33); near its bottom are two parallel lines.

W 35 Side view of the hilt (picure W 33). The grip cap is provided with tabs, the guard has a folding hinge without holes. (HMR)

Description:

Blade: Stiff-backed blade with brace. No decorations or maker's name to be found.

Hilt: The brass hilt consists of a half-basket. In its upper part (proceeding from the handle cap), a large unclear anchor stretches to the middle of it. The outer rim of the basket is defined with decorative lines which—in the lower part of the basket—end at the guard in the form of a sea serpent.

The grip cap is decorated with an eagle's head.[54] The eagle's feathers cover the cap only in its uppermost part. Cap tabs are present.

Technical Data of examined Schleswig-Holstein naval officers' sabers of 1849:		
Weapon	1	2
Length sheathed	930 mm	
Length drawn	940 mm	918 mm
Blade shape with guard	triangular, with guard	triangular,
Blade length	780 mm	791 mm
Blade width	29 mm	
Blade thickness	8 mm	
Hilt height	127 mm	
Bow shape	C-shape	C-shape
Handle cap	Eagle's head	Eagle's head
Handle grooves	10	14
Slit/holes	none	none
Lock	none	none

The inner, or quartside, of the basket is fitted with a snap and thus removable.

The grip ring is left smooth.

The grip itself consists of a wooden core covered with fish skin. Ten to fourteen grooves hold the wire wrapping. This consists of two entwined brass wires.

Scabbard: The leather scabbard has three brass fittings. Some are smooth, while some have indentations at their ends.

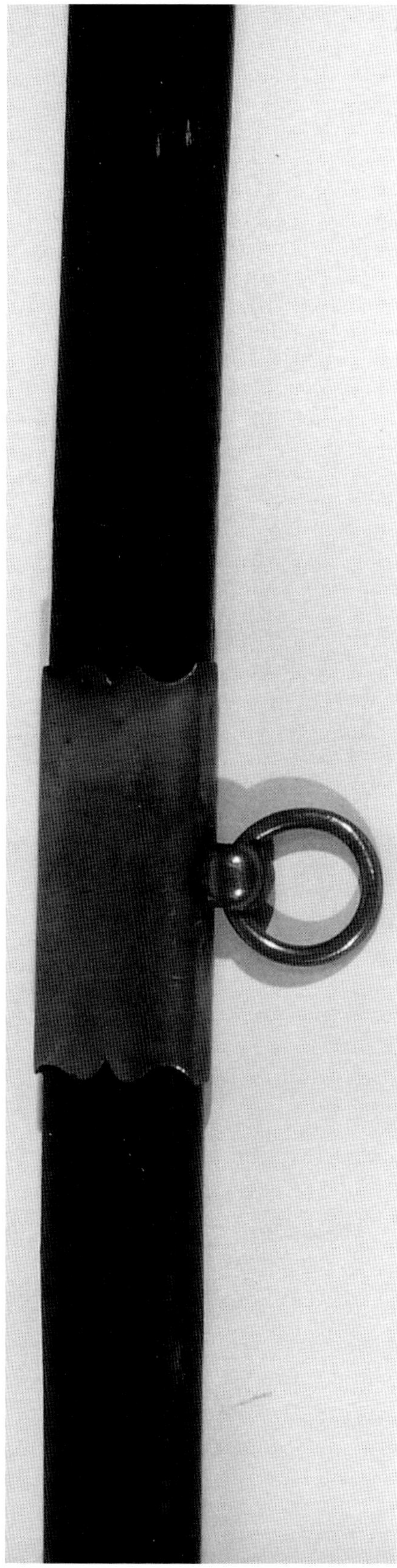

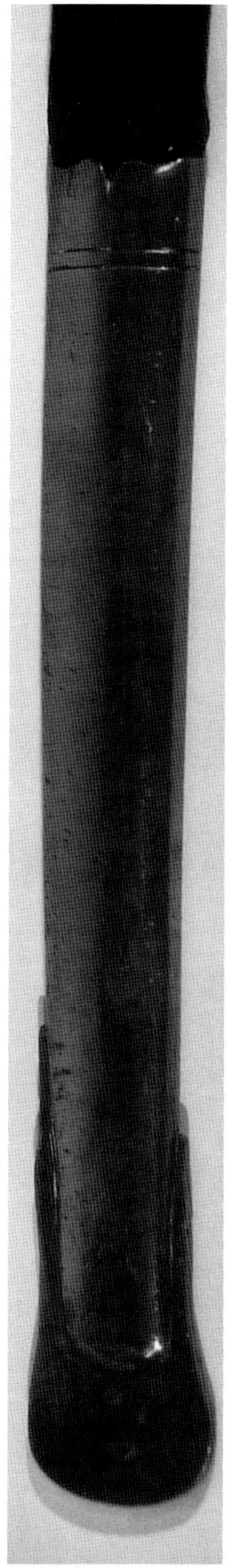

Left: W 38 Middle scabbard fitting of the Schleswig-Holstein naval saber (picture W 33) with no decorative lines. (HMR)

Right: W 39 Lower scabbard fitting of the Schleswig-Holstein naval saber (picture W 33) with two parallel lines in its upper area. (HMR)

The Navy of the North German Federation (1867-1870)

After the 1866 war, the states and cities north of the Main founded the North German Federation. In its constitution, the Prussian Navy became that of the North German Federation. Command and administration, though, remained under Prussian authority. In Article 53 of the constitution, which came into power on July 1, 1867, the organization of the navy is defined:

The Federation Navy is unified under Prussian command. The organization and composition of it are authorized by His Majesty the King of Prussia.... The necessary funds for the foundation and maintenance of the fleet and the institutions that go with it will be provided by the Federation's treasury.

As for the naval saber, it can be stated that no official changes are known prior to the foundation of the German Empire in 1871. In historical literature about the North German Federal Navy it is only said: *The weapon is the curved naval saber in a leather scabbard, sashes do not exist.*[55] Heinrici[56] also stated briefly in 1868: *Naval officer's saber, leather scabbard with brass fittings and ivory handle.*

Hand drawing of the 1849 naval saber (picture 14) from the files of the Imperial Admiralty (process dated June 22, 1872); the old Prussian/North German lion's-head shape and the grip with seven grooves are easy to recognize, likewise the prescribed knotting of the portepee by means of a woven linen strap on the hilt of the weapon.[57]

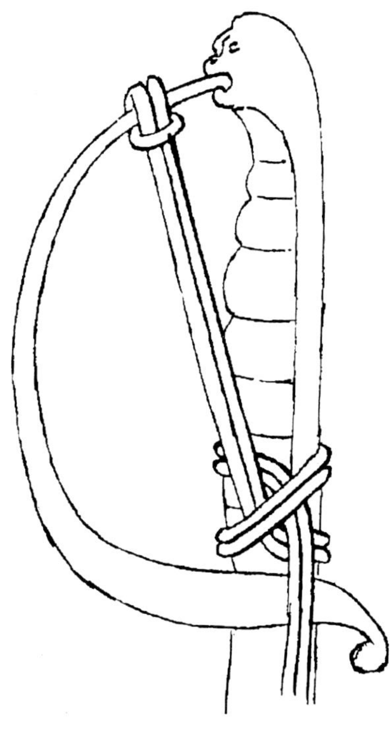

14 Hand drawing of the 1849 naval saber. (BA-MA)

The Navy of the German Empire (1871-1918)

On April 16, 1871, the constitution of the German Empire, which was founded on January 18, 1871, came into force. In articles 53 to 55 the regulations concerning the navy and shipping were codified. Of the navy it was said: *The navy of the Empire is unified under the command of the Kaiser. Its organization and composition are up to the Kaiser, who appoints the officers and administrators of the Navy, and to whom they, along with the men, are to swear an oath.... The necessary funds for the formation and maintenance of the fleet and the institutions connected to it will be provided by the Imperial Treasury.*[58] From this point on, the navy held a special position among the armed forces: in war and peace it was subordinate to a unified high command, that of the German Emperor. The kings of Bavaria, Saxony, and Württemberg had special rights over the troops from their lands, which need not be gone into at this point.

The first regulations for changes in uniforming and arming naval members were made in 1872. For example, the Chief of the Admiralty ordered on July 9, 1873, that naval officers wear a six-pointed star above the stripes on their sleeves. Finally, on February 25, 1874, the Chief of the Admiralty published the uniform regulations ordered by the Kaiser on February 24, 1874, applying to all groups of naval personnel except the naval garrison administrative and lazarette personnel. In this order it was said of sidearms:

13. Saber. Naval saber in black leather scabbard with gilded fittings, gilded bow and ivory handle, on a belt of matte black leather with round side straps, gilded buckle and carbine hook.

15 Sea Captain Conrad Dietert with an early imperial saber for sea officers. MSM Archives

16 Details of Sea Captain Dietert's saber.

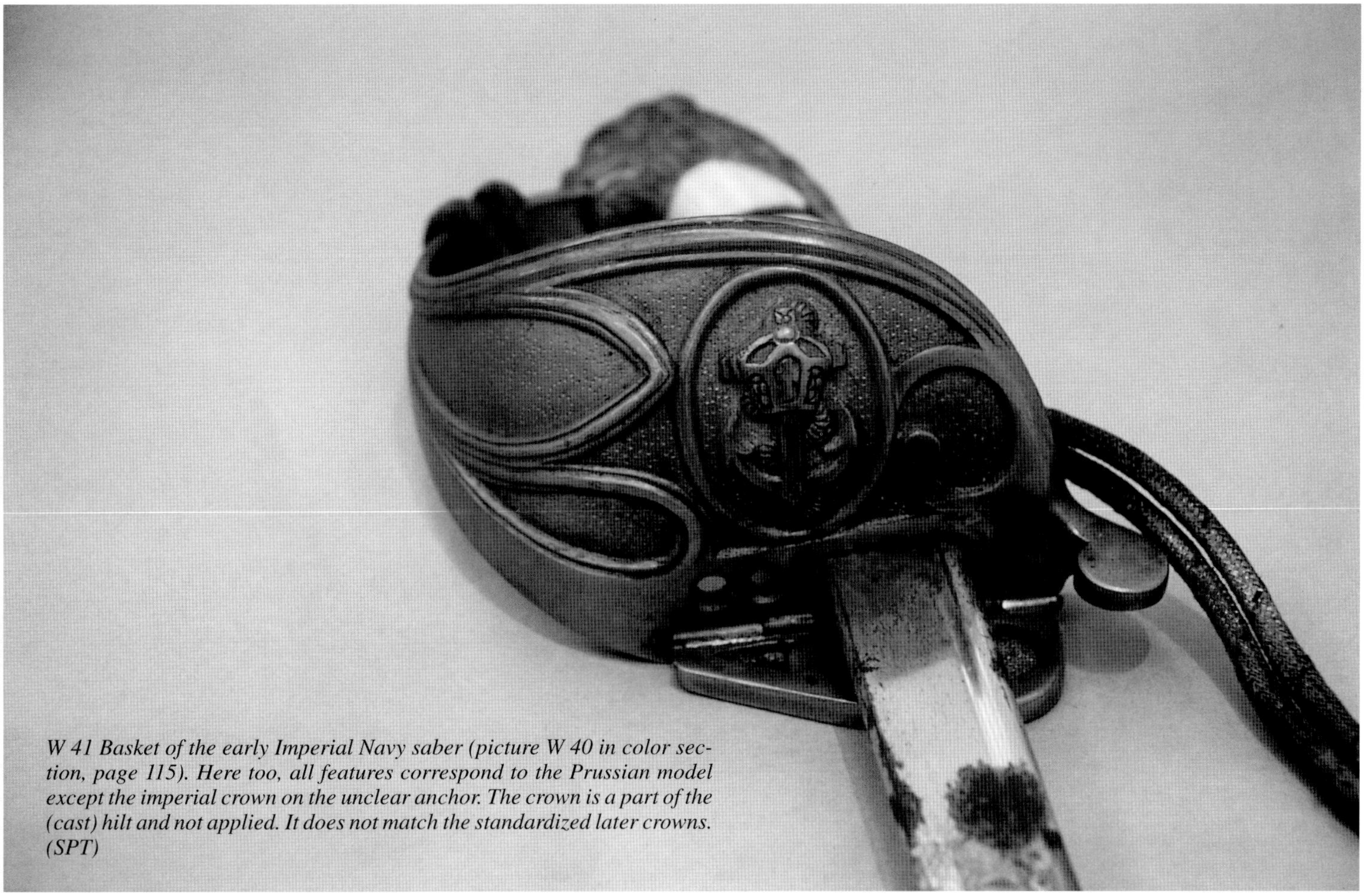

W 41 Basket of the early Imperial Navy saber (picture W 40 in color section, page 115). Here too, all features correspond to the Prussian model except the imperial crown on the unclear anchor. The crown is a part of the (cast) hilt and not applied. It does not match the standardized later crowns. (SPT)

With it were to be worn:

14. Portepee. Of silver worked with black and red silk, on two round silver cords, worked with black and red, which are united.[59]

With these regulations, a definite model had been prescribed for all members of the Imperial Navy who had the right to wear a saber, though obviously one—essential—change of this model from the earlier saber worn by officers of the Royal Prussian Navy was not cited.[60] But this will come up later.

One small but not insignificant change in uniforming was ordered on August 17, 1875, when a new form of buttons for officers' clothing was introduced. The buttons were now to have an imperial crown over an unclear anchor.[61]

Some eight months later, the aforementioned uniform regulations for the officers etc. of the Imperial Navy were modified again and defined more precisely. In an order of the Chief of the Admiralty, dated April 7, 1876, it was stated as to #13, re the sidearm:

The naval saber is slightly curved; the gilded bow is not cut through and bears on its lower external side an unclear anchor with an imperial crown, according to the pattern ordered in Naval Instruction Sheet of 1875 No. 16, page 142.

It is notable that only some five years after the founding of the Imperial Navy was a changed model of the naval saber officially introduced. The essentially new feature of the weapon was the hilt, which now bore the imperial crown over the unclear anchor.

This regulation, of course, does not rule out the possibility that in rare cases naval sabers with the Imperial crown on the hilt already existed before the date of the order.[62]

At this point, a small incident that concerns the Imperial Navy's two saber models should be reported. The Navy was required (probably after 1878) to order needed spare parts for naval officers' sabers from the armament factory in Erfurt.[63] There, though, the orders could not be filled, on account of the differences of the weapons and the lack of a (dependable) model weapon. This situation was documented in a letter of November 28, 1883, from the Admiralty in Berlin to the Inspection of Naval Artillery in Kiel, which was responsible at that time for the navy's weapons. The letter in question is quoted here:

According to a communication from the Ministry of War, various orders were made by naval units from the armament factory in Erfurt for replacement parts for naval officers' sabers, the filling of which creates problems, in that the factory is not in possession of a model of this saber, and the required model pieces required from the naval units in the individual cases differ from each other in form and dimensions.

Regarding the establishment of a uniform form of this weapon, I most humbly request of the recipient the prompt sending of two naval officers' sabers, to be obtained from the Naval Artillery Depot in Friedrichsort, that correspond to the uniform regulations for the officers etc. of the Imperial Navy stated in Paragraphs 1 to 13 of M.V.Bl. for 1874, pp. 27ff., as well as in the order of April 7, 1876, M.V.Bl. p. 69.

The cost of the two sabers is to be covered according to Kap. 61 Tit. 6.

The Chief of the Admiralty, von Capelle

From this letter it can be learned that from 1873 to 1883 two models existed side by side in the Imperial Navy. In addition, only after the conclusion of the cited process in 1883, or even in 1884, could the saber model introduced by the order of April 7, 1876 (unclear anchor with Imperial crown), finally be established as the weapon of the sea officers. This model was officially retained until the end of the Imperial German Empire. In the *Clothing Regulations for Sea Officers* etc. of March 25, 1890, February 28, 1898, and October 29, 1900, the regulations are found almost word for word, so that further quotations are not needed.

It should not remain unnoted that through the further development of the navy new groups of personnel received authorization to wear the naval saber. For example, the armorers of the Imperial Navy, by A.K.O. of October 22, 1878, as long as they had completed fifteen years of service with blameless conduct and loyal performance of their duties.

The Saber of the Sea Officers of 1874

First of all, it should be made clear how the naval saber specified in February 1874 appeared. One can look in vain through old documents or the older literature for detailed information on the appearance of the naval saber, especially for its dimensions.

In November 1875 Adalbert Mila wrote about the sidearm of the sea officers etc., though crudely:[64] *Naval sabers in black leather scabbards with gold fittings, gilded hilt, ivory grip on a black leather hanging strap, the belt of which is worn under the vest.*

On it the silver naval portepee worked with black and red silk, the latter differs from the portepee of the land army in that the tassel does not hang on a band, but on two round silver cords worked with black and red.

According to this description, the main change in the weapon is that now the bow of the hilt is gilded, whereas the bow of the previous weapon—as noted—at least officially was simply made of brass.

17 Paince Heinrich of Prussia, here as a Sea Captain, with the earlier saber for sea officers; note the anchor without a crown. C.S. Archives

W 42 Saber for sea officers of 1874. It differs from the Prussian naval saber in the new shape of the lion's head, especially the greater length of the mouth. (PS)

W 44 Basket of an early saber (picture W 43 in color section, page 118). The form of the basket still corresponds to the 1849 model. (PS)

In addition, it is noteworthy that the Imperial Navy's first saber obviously did not yet have a crown over the anchor, which otherwise would have been mentioned as a particular detail. In other words, the new sidearm was very similar to the Prussian naval saber. On the other hand, we can assume that the Royal Prussian model was "determined"—as was the official term—anew for the Imperial Navy.[65] The assertion that the Prussian saber model of 1849 was "simply continued to be worn" was not based correctly on the sources. Otherwise, in the uniform regulations there would very probably have been a reference to the old model. Then too, the Admiralty would not have referred to two (new) models in the quoted sources (and cited the applicable regulations).

The fact that there was no crown on the older model also explains the phenomenon that is seen by the observer of photos of young naval officers. Numerous officers and administrators actually wore a saber without an Imperial crown! The presently best-known personality who often wore such a "crownless" saber was *Prince Heinrich of Prussia,* the brother of *Kaiser Wilhelm II.* At various times and various ranks, and finally as Grand Admiral and Inspector-General of the Navy, he was still photographed with this type of saber. Prince Heinrich entered the Navy in 1877 and accordingly could only have worn the model with the Imperial crown, which was prescribed as of 1876, if the two models had not existed officially side by side.

Description:

Because of the similarity to the old Prussian saber model, a detailed description will be omitted.

W 45b The lion's head of an 1849 naval saber (compare with W 45a). (PS)

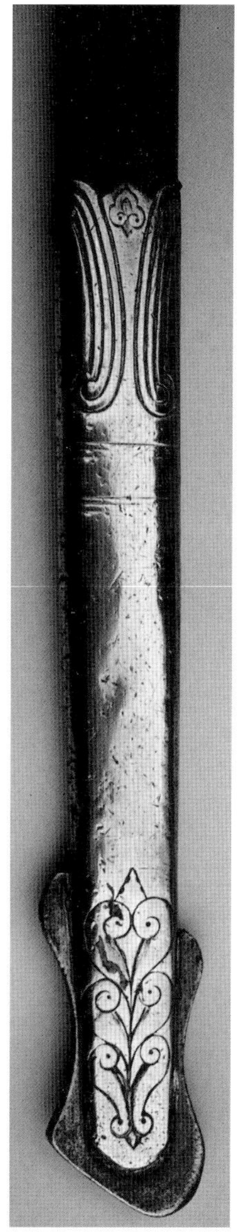

W 48 The lower scabbard fitting of the saber. (PS)

W 46 The upper scabbard fitting of the 1874 naval officers' saber. (PS)

W 47 The middle scabbard fitting of the saber. (PS)

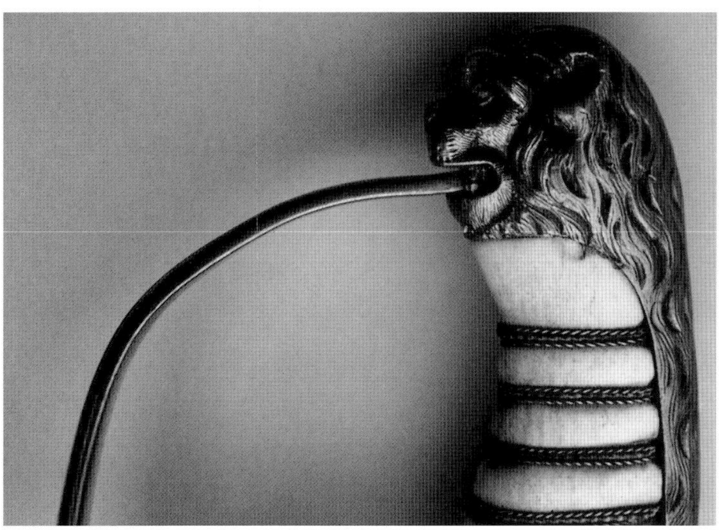

The quotation from Mila's book, though, saying that the hilt was now gilded, is noteworthy.

On the basis of various photos it can be assumed—contingent on final proof—that the naval saber accepted in 1874 had a different, more timely lion's head than the Prussian type. Where the "Prussian lions' heads" were mostly rounded, though the size of the actual head varied, and had short, almost closed mouths, the new head already resembles very closely the heads on the most recent imperial sabers. The extent of the mouth now extends for almost half of the lion's head. In comparison it can be noted that the lion's mouth of the old Prussian model extended over only about a third of the head. The clearest mark of the newer heads, though, is the wide open mouth, with the teeth usually clearly visible.

W 45a The lion's head of an 1874 naval saber (compare with W 45b). (PS)

Technical Data of examined 1874 Imperial Navy sabers			
Weapon	1	2	3
Length sheathed	854 mm	918 mm	915 mm
Length drawn	839 mm	910 mm	910 mm
Blade shape	triangular, reinforced	triangular, reinforced	triangular, reinforced
Blade length	713 mm	779 mm	783 mm
Blade width	23 mm	25 mm	25 mm
Blade thickness	7 mm	7 mm	7 mm
Hilt height	130 mm	129 mm	120 mm
Bow shape	angular	angular	angular
Slit/holes	none	none	none
Attacher	yes	yes	yes
Handle grooves	7	7	8

The Saber for Sea Officers of 1876

As we know, as of April 7, 1876, a new model of the naval saber was introduced. In the fourth edition of his uniform book of 1881, the already quoted Adalbert Mila wrote:[66]

Naval saber in black leather scabbard with gilded fittings, gilded bow and ivory handle, on black leather strap, the belt worn under the vest. The straps are not wide as in the Army, but round and linked with the rings by carbine hooks.

The bow is not perforated and bears on its lower outside an unclear anchor with the Imperial crown. On it is the silver naval portepee worked with black and red silk, the latter differs from the Army officers' portepee in that the tassel does not hang on a band, but on two round silver cords worked with black and red.

Description:

Blade: As with the Prussian saber, the Imperial pendant regularly has a reinforced back. The tip of the blade is sharpened at the back. The blade, like those of Prussian weapons, is decorated with etched floral and maritime designs. In the middle of the decoration is an unclear anchor, at first without, and later with an Imperial crown.

Hilt: The now gilded hilt consists of a three-quarter basket with raised line decor, in the middle of which is an oval with an unclear anchor under an Imperial crown.

The inside of the basket is equipped with a hinge and therefore folds. The weapons regularly had a hole in the flap, which was able to hold a spur soldered onto the mouth fitting of the scabbard.

18 Kaiser Wilhelm II in an admiral's uniform, with the typical 1876 naval officers' saber. (MOV Archives)

The form of the bow is that of a rounded rectangle.

The grip cap is made with a lion's head. The lion's mane covers the entire cap down to the handle ring. The lions' heads apparently did not have red/green eyes in the early models.

The grip ring is decorated with channelings and/or raised rings.

The grip itself is made of ivory or bone. Twelve or thirteen grooves hold the wire wrapping. This usually consists of three plaited brass wires.

Scabbard: The scabbard is made of black leather and has three yellow-metal fittings. On the mouth or middle fitting there is always a blunt soldered ring eyelet.

The decorative grooves of the middle fitting no longer extend regularly over its entire length. The vertically running decorative flutings are usually divided into two areas by horizontal grooves.

The mouth fitting is asymmetrical.

A more exhaustive description will be omitted. Details and particulars of the 1876 naval saber are provided in the next chapter.

W 63 The 1876 model naval saber. From the drawn and colored pattern book—from the early sixties of the 19th century—by Carl Eickhorn of Solingen. This is the oldest known picture of a naval saber in a pattern book. (DKS)

W 65 Basket of the earlier naval saber (picture W 64 in color section, p. 115). Here too, the flat tab still shows much similarity with the Prussian model. The Imperial crown has multiple shields at the bottom. This is the early (first) Imperial crown. The wrapping of the anchor chain is very different from that of ordnance weapons. It first runs behind the right part of the crossbar, is then carried by the anchor shaft in the middle, and is then brought forward over the left part of the crossbar. Then it falls away, finally brought forward from in back of the right anchor fluke. (PS)

W 66 The basket of another early naval saber. This Imperial crown too, though clearly wider, has multiple shields at the bottom. (PS)

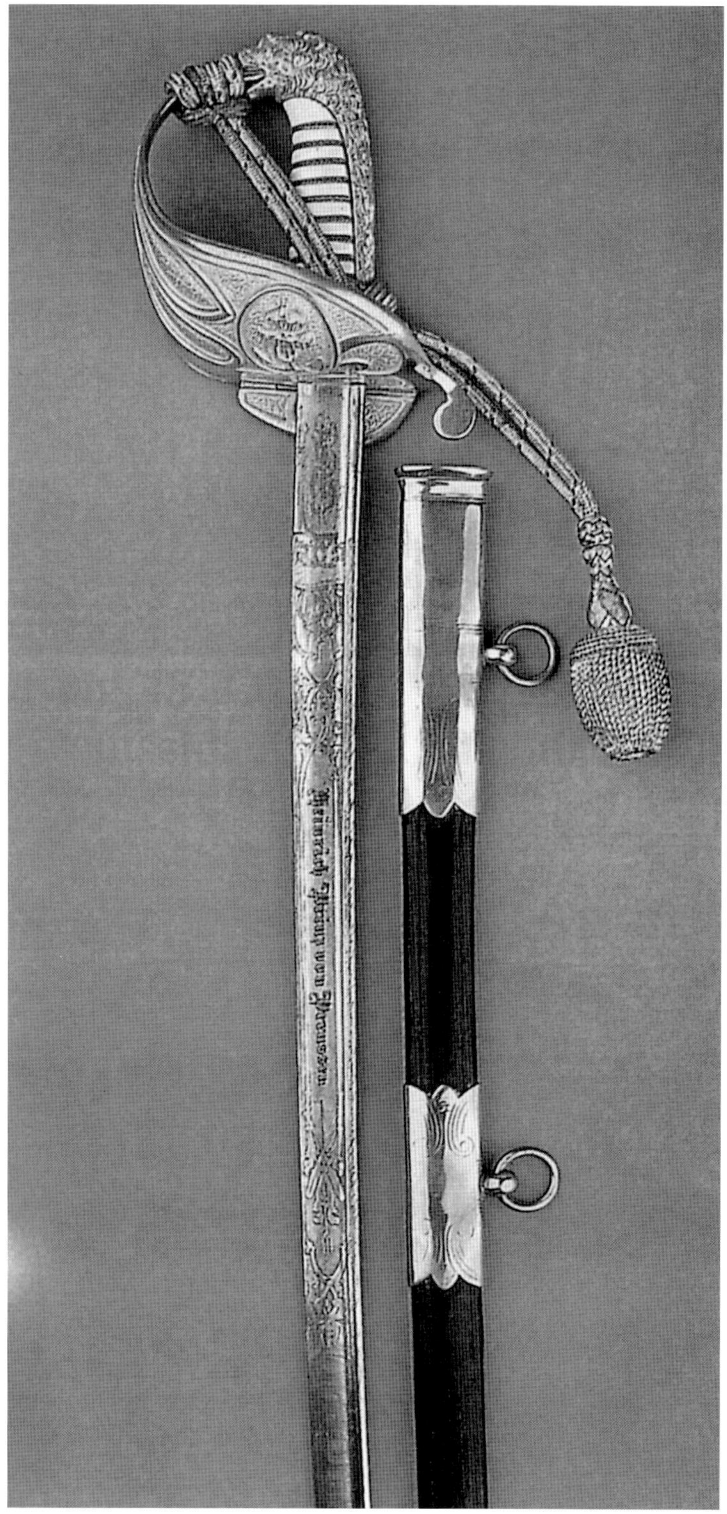

W 67 One of the sabers of Prince Heinrich of Prussia. The design of the medallion on the basket is like that of the previous weapons, particularly the saber in picture W 63 (p. 51). The weapon can be dated by the inscription on the blade: 1st October 1882. The saber was probably presented to Prince Heinrich of Prussia by his father, Kaiser Friedrich III, for his service on SMS Olga. This ornate saber was obtained from A. Wiener of Berlin. (STJ)

The Ordnance Weapons

The weapons created by the military for, among others, the lower officers with portepee,[67] are also known as ordnance sabers.[68] These sabers will be described first before we look at the variants created by weapons merchants and uniform tailors, for these sabers are supposed to correspond exactly to the regulations of the Imperial era.

The ordnance weapons known to date all correspond to the 1876 model. To be sure, it can be noted that even the ordnance weapons differ in minor details. The often-cited uniformity in clothing and sidearms (at least for officers and portepee underofficers) existed neither in the Prussian Army nor the Imperial Navy!

19 Portepee lower officer with the typical naval saber, the portepee tied as prescribed. SPT Archives

20 A deck officer with the Imperial naval saber. D.S. Archives

W 49 The hilt of a service-provided naval officer's saber (1876 model) for portepee lower officers, so-called ordnance weapon, from the mid-eighties of the 19th century. (PS)

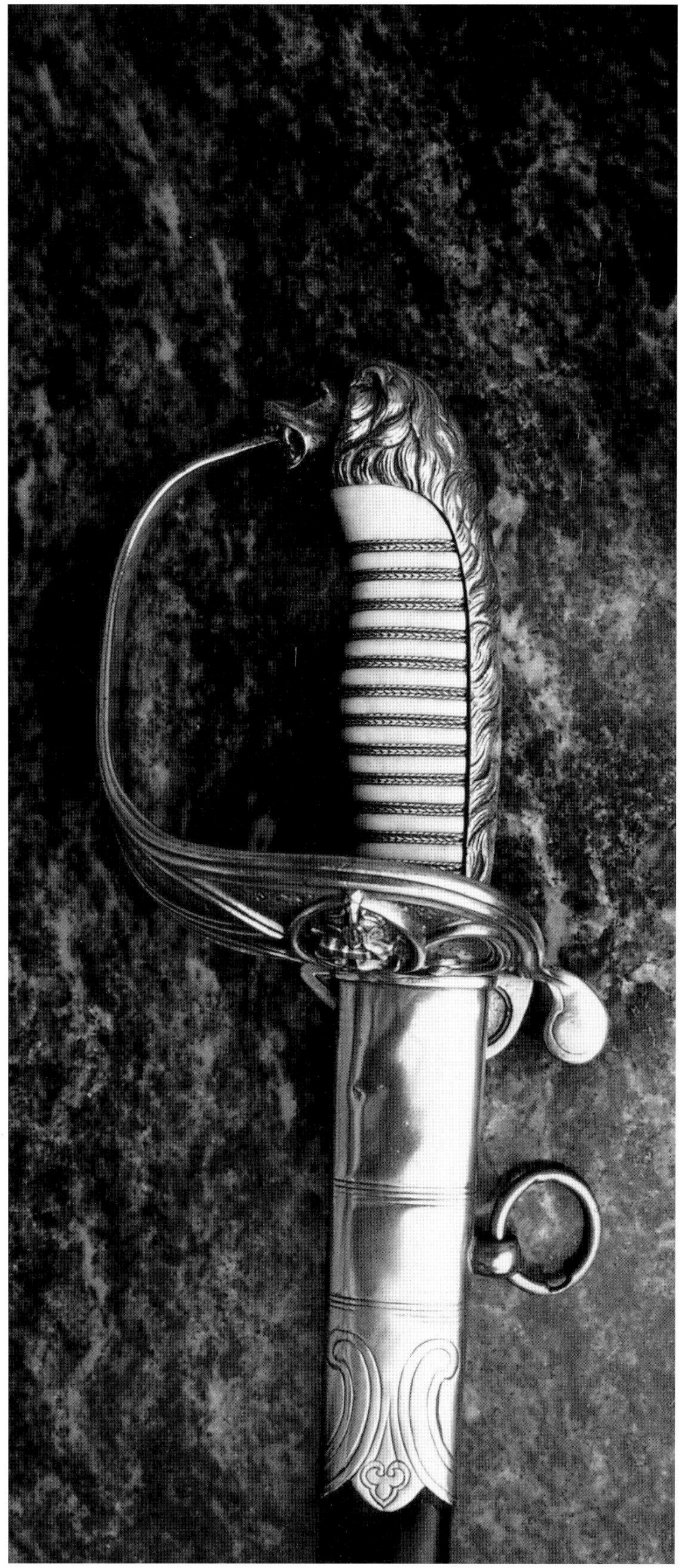

W 50 The hilt of another service-provided naval officer's saber, made after 1890. (PS)

W 51 The basket of a saber from the 1880s (picture W 49). The anchor chain is at first behind the right part (seen from above) of the crossbar and wrapped around the anchor shaft only in the middle of the anchor, then running behind the right anchor fluke. The tab to the right of the medallion is decorated with a groove.

Production: In the Prussian Army the Feldwebel, Wachtmeister, and Oberfeuerwerker were allowed, since June 18, 1822, to wear the officers' sidearm. According to an order of July 6, 1822, the applicable soldiers were required to provide the weapon from their own means. Only by order of June 7, 1858, was it allowed, after numerous requests, to have the weapon bought by the troop unit, what with the meager financial means of the applicable personnel, to pay for the weapons officially from the Weapon Repair Funs. They were to be ordered directly from the Solingen weapons factories or the applicable Weapon Revision Commission.

Only by order of March 6, 1876, was it determined that the sidearms were to be obtained and delivered to equip the troop units in the same way and on the same principles as the others.[69] The order of December 8, 1877, required that as of January 1, 1878, all sidearm components were to be ordered from the gun factory in Erfurt.[70] These requirements presumably also applied to the Navy; nothing else can be proved. By decree of August 5, 1884, the Chief of the Admiralty made known that the purchase of spare parts for officers' sidearms from the gun factory could continue, but as of October 1, 1884, the purchase of spare parts of crewmen's weapons and complete officers' sidearms should be done from specific artillery depots.[71]

W 52 The basket of the newer saber (picture W 50). The anchor chain is wound in a more complicated way. The Imperial crown, probably according to an order of December 6, 1888, is fitted with—barely visible—bands, the "Pendilien." (PS)

Markings and Stampings:

The weapons for the men of the Prussian Army, as well as the Navy, were subjected to quality control—the so-called Revision—before they were accepted from the manufacturers by the appropriate agencies. As a sign of successful testing, the individual parts of the weapon had the so-called acceptance symbol (a Gothic majuscule with a crown at the top) stamped on them with a punch. The letter is the initial of the responsible weapons tester's name. The Army testers—as an official service—also tested the naval sabers. The weapons for the naval portepee lower officers were, at least from 1876 on, also subjected to this quality control and stamped appropriately.

On the back of the blade, at the height of the dull edge, some, but not all, ordnance weapons also bear the so-called military fiscal stamp; this is a **W** under a (royal) crown, with a two-digit number, such as 85, under it. This stamp means that the weapon was taken into military possession in the reign of the Kaiser and King **W**ilhelm (I) in the year 18**85**.

W 53 The underside of the guard of the later naval saber (picture W 50, p. 54) with the acceptance stamp in the form of a Gothic majuscule with a crown. The letter is the initial of the testing official's name. (PS)

W 55 An acceptance stamp on the back of the blade of tne newer naval officer's saber (picture W 50, p. 54). At that time the military fiscal stamp was not used. (PS)

W 54 On the back of the blade (naval saber in picture W 51, p. 53), the crowned military fiscal stamp W 85 can be seen. This stamp was applied by a Royal Prussian acceptance official. The W stands for the ruler, King Wilhelm, who as German Kaiser was simultaneously the supreme commander of the Navy. The number 85 indicates the year of acceptance in the military budget, 1885. Under it is an acceptance stamp. (PS)

W 56 The stamp of the Royal Prussian Artillery Depot in Erfurt. In the latter half of the 19th century, this depot was responsible for supplying the Navy with hand weapons. (PS)

Sometimes on the terzside of the dull edge the name ERFURT is stamped under a (royal) crown. In Erfurt was the depot from which the naval units also were supposed to order their replacement weapons and weapon parts.[72]

The Navy also applied its own fiscal stamp to show that the weapon belonged to its supplies. In a directive from the Chief of the Admirality dated March 29, 1882, the new stamp in effect as of April 1, 1882, was described as follows: *Stamp with the letter M and Imperial crown over it.* The letters **K.M.**, stamped previously, for **K**önigliche-**K**aiserliche **M**arine, were dropped.[73] The naval fiscal stamp was applied to naval officers' sabers on the quartside of the end button at the parrying bar. What with the military and naval fiscal stamps and the acceptance stamp, the weapons were and are easily identified as to when they were accepted and issued for service.

W 57 An early naval fiscal stamp: K M (saber in picture W 51, p. 55). The two letters stand for Kaiserliche Marine (Imperial Navy). The naval unit stamp II M D 145 means Second Seamen's Division, weapon number 155. (PS)

W 58 An example of the naval fiscal stamp used as of 1882, a crowned "M" standing for "Marine" (Navy). The unit stamp II M D 67 stands for II. Seamen's Division, weapon no. 67. (PS)

The ordnance weapons were also provided regularly with a stamp, which gives the present-day enthusiast informatiuon about their use by a naval or troop unit.[74]

For the Prussian Army there was an *Instruction for the stamping of hand weapons* etc., which was updated and reprinted in 1873, 1877, 1881, 1890 (with a new printing in 1893), 1897 (with additions in 1899), and 1909. The last edition also verifies the troop stamp: *The stamping shall essentially only serve to make the weapons easily recognizable to the individual man while they are in use. It does not have the purpose of making the individual troop unit recognizable for weapons lost in war or put into weapon collections.*

An inclusive instruction manual just for the Navy, dealing especially with the naval-unit stampings, is not known. For stamping naval officers' sabers, appropriate stampings used by the Prussian Army were resorted to, or the Navy's own stamping directives, which actually regulated stampings for other areas. Such stamps are found on the quartside flap of naval sabers and the mouth fitting of their scabbards. One example would be the directive of the Chief of the Admiralty about the *Stamping of the leather pieces that belong with hand weapons in the seamen's and shipyard divisions and in the naval cadets' unit* of February 25, 1877, which says:

The rifle straps, visor flaps, cartridge cases, body belts and hunters' pouches are as of now with the abbreviated designation of the naval branch:

I M D (First Seamen's Division),
II M D (Second Seamen's Division), and
S J A (Naval Cadets' Unit)

and with a number beginning with 1 for every naval branch... to be stamped.

Notice the lack of periods after the numbers and letters.

W 60 The naval stamp I W D 32 means First Shipyard Division, weapon no. 32. (PS)

W 59 The naval stamp I. M. A. A. 17 means First Seamen's Artillery Unit, weapon no. 17. (PS)

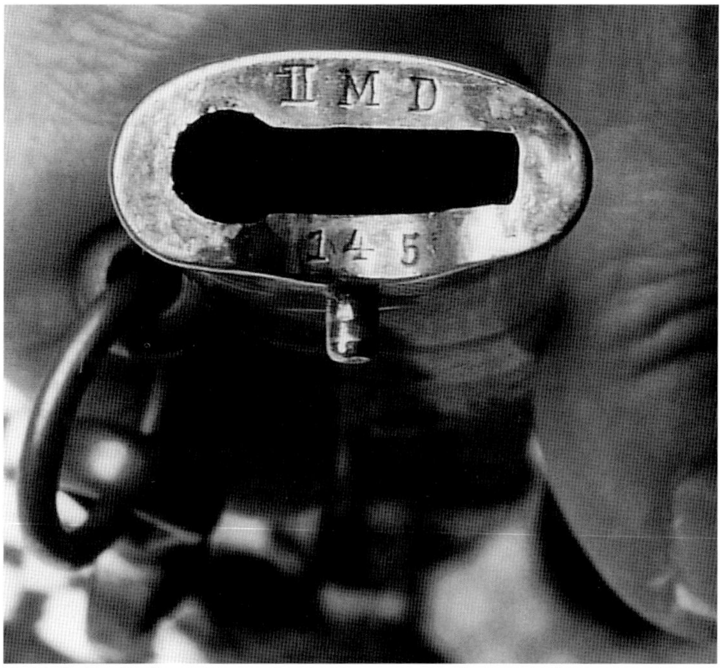

W 61 Naval unit stamp on the scabbard (compare picture W 57), here on the covering plate of the mouth fitting. (PS)

W 62 Naval unit stamp, here on the quartside of the mouth fitting (compare picture W 60). (PS)

On the other hand, naval unit stamps with periods were also called for. Thus, in the *Instructions in regard to the hunting rifle m/71 (Stamping,...)* of June 22, 1875, it was stated:

...3) For the designation ad a (meaning the names of the naval units) *the following abbreviations are to be used:*

I. M. D. (First Seamen's Division)
II. M. D. (Second Seamen's Division)
S.J.A. (Naval Cadets' Unit)
I. S. B. (First Sea Battalion)
I. S. A. A. (First Sea Artillery Unit)
II. S. A. A. (Second Sea Artillery Unit)

4) Every seamen's division and the cadets' unit uses its own consecutive numbers beginning with 1 on the m/71 hunting rifle....

Both types of naval stampings—with and without periods—are also found on the ordnance weapons of the sergeants. According to present-day observation it can thus be assumed that the stamps of the older weapons are without periods and those of the newer ones usually with them.

Variations:

It was customary that when ordnance weapons were being manufactured, a stamped specimen of the weapon was—at least sometimes—made available to the manufacturers for copying. In addition there were dimension tables and later measuring tables, which prescribed all measurements and materials of the sidearms in detail. This information scarcely left the manufacturers any latitude in production if they did not want to run the risk of having their weapons rejected by the Revision.

It must, though, be made clear in the case of naval officers' service sabers that there are nevertheless more variations than in other ordnance weapons of the Army. In this respect we can refer again to the already quoted Admiralty directive of November 28, 1883. Differences will be shown in the next table.

Notes on the chart at right:

2. In addition to the acceptance stamps, the weapon is also marked with the military fiscal stamp W 85 and the ERFURT stamp.
3. This weapon has acceptance stamps on all parts, but no military fiscal stamp.

Technical Data of examined Imperial Navy sabers (ordnance weapons)

Weapon	1	2	3	4
Length sheathed	935 mm	870 mm	940 mm	1005 mm
Length drawn	910 mm	861 mm	925 mm	967 mm
Blade shape	Triangular, reinforced	Triangular, reinforced	Triangular, reinforced	Triangular, reinforced
Blade length	773 mm	725 mm	786 mm	832 mm
Blade width	24 mm	25 mm	26 mm	25.5 mm
Blade thickness	6.1 mm	2.5 mm	5 mm	6 mm
Hilt height	137 mm	136 mm	139 mm	135 mm
Bow shape	angular	angular	angular	angular
Crown in oval	no bands	no bands	with bands	with bands
Attacher	yes	yes	yes	yes
Grip grooves	12	12	12	12
Fiscal stamp	K.M.	M + crown	M + crown	M + crown

(Listed in chronological order of their acceptance by the Navy.)

4. This weapon has no acceptance or fiscal stamps anywhere. The form of the hilt also has clear differences—made thus by the manufacturer?

- The form of the Imperial crown and the anchor on the baskets of sabers changed in the course of time (see the pictures). The length of the anchor and its chain also vary.
- On newer weapons, very probably because of the A.K.O. of December 6, 1888,[75] there are bands, the so-called Pendilien, which occur under the Imperial crown.
- The grip ring of older weapons is decorated with designs of reed leaves, the newer ones with oak leaves.
- Obviously, the standards for acceptance of ordnance weapons for portepee lower officers—as already noted—were rather generous.

Privately Made Weapons

A great number of manufacturers made sabers for naval personnel. The capable Solingen firms produced, according to taste and financial considerations, whatever was liked and would sell. Variations, it seems, exist in Imperial Navy sabers in almost as great numbers as there were officers, deck officers, naval administrators, and the like. *For differences in the cut of the coats, shapes of caps, ties and shirt collars,* ***length and width of sabers and daggers etc.*** *style ruled more or less capriciously and boldly,* thus reports a contemporary who was also a sea officer (*Admiral Hopmann).*[76]

21 Leutnant zur See Wahrendorff (circa 1874) with a naval officer's saber that has a straight blade and scabbard. MSM Archives

Below: 22 Leutnant zur See d. R. Georg Sachau (circa 1917) has a saber with basket flap. DS Archives

Right: 23 Close-up of a naval officer's saber with lavish floral decorations. MSM Archives

Below Right: 24 Close-up of a naval officer's saber with basket flap. Note the special decorations on the scabbard fittings. They are also uncommonly long. MOV Archives

26 A deck officer wearing a saber with trapezoid-shaped flap basket. Huss Archives

25 A deck officer wearing a saber with lavishly decorated oval flap basket. D.S. Archives

Describing all variations fully and showing pictures of them could not be done within the limitations of one book. To offer an overview of the possible variations, nevertheless, the most important possibilities of variation were gathered by category. Differences in lengths were not included. In selecting the pictures, the authors try to afford an understandable overview of what was manufactured and worn.

27 Kapitän zur See von der Groeben, with an unscrewable, demountable saber. Note the small ball-shaped screw. MSM Archives

28 A sea officer with a nonstandard naval officer's saber; the hilt has a pierced basket. D.S. Archives

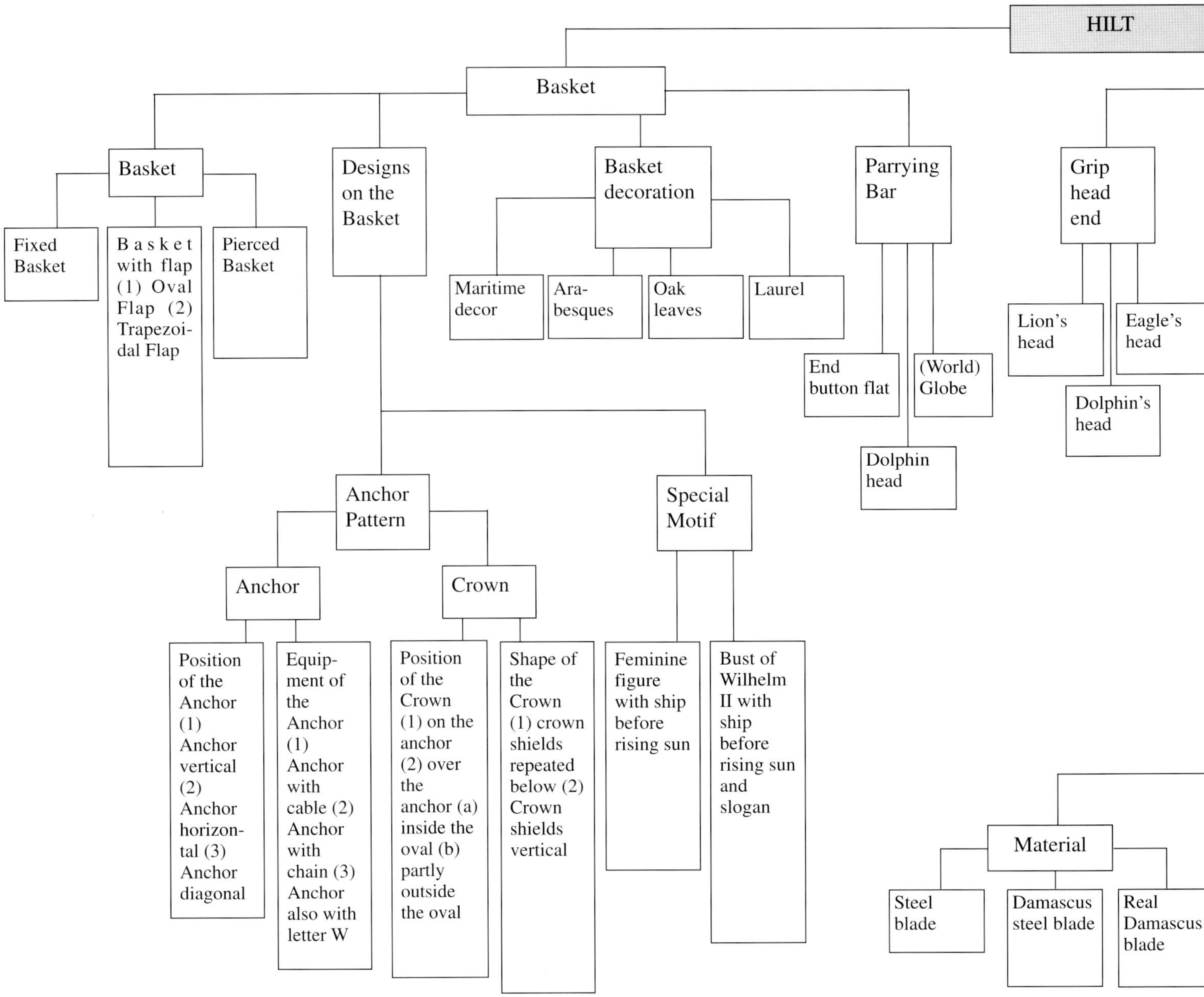

Diese Übersicht erhebt keinen Anspruch auf Vollständigkeit.

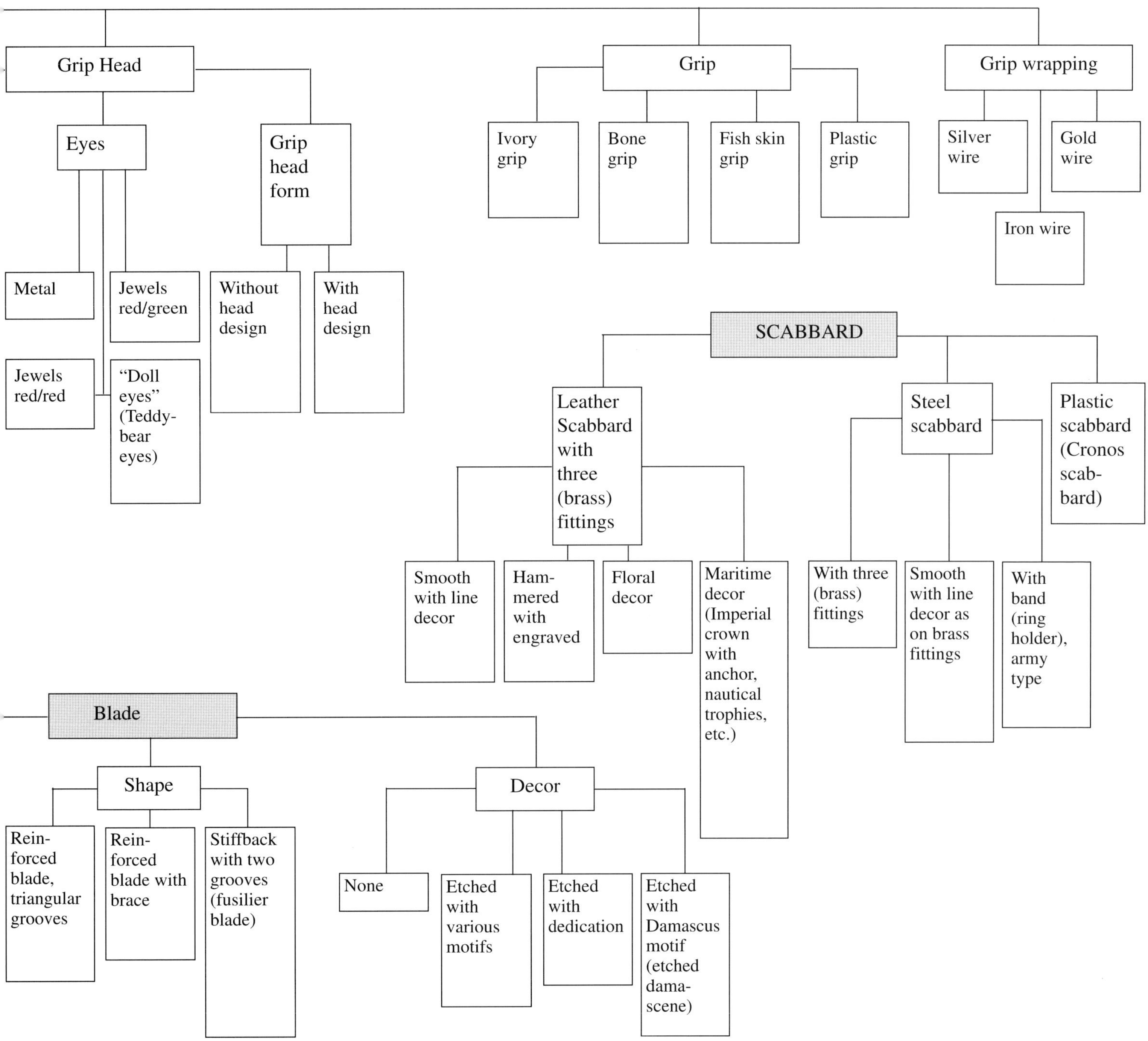
Grip Head
Eyes
Grip head form
Metal
Jewels red/green
Jewels red/red
"Doll eyes" (Teddy-bear eyes)
Without head design
With head design
Grip
Ivory grip
Bone grip
Fish skin grip
Plastic grip
Grip wrapping
Silver wire
Gold wire
Iron wire
SCABBARD
Leather Scabbard with three (brass) fittings
Smooth with line decor
Ham-mered with engraved
Floral decor
Maritime decor (Imperial crown with anchor, nautical trophies, etc.)
Steel scabbard
With three (brass) fittings
Smooth with line decor as on brass fittings
With band (ring holder), army type
Plastic scabbard (Cronos scab-bard)
Blade
Shape
Rein-forced blade, triangular grooves
Rein-forced blade with brace
Stiffback with two grooves (fusilier blade)
Decor
None
Etched with various motifs
Etched with dedication
Etched with Damascus motif (etched dama-scene)

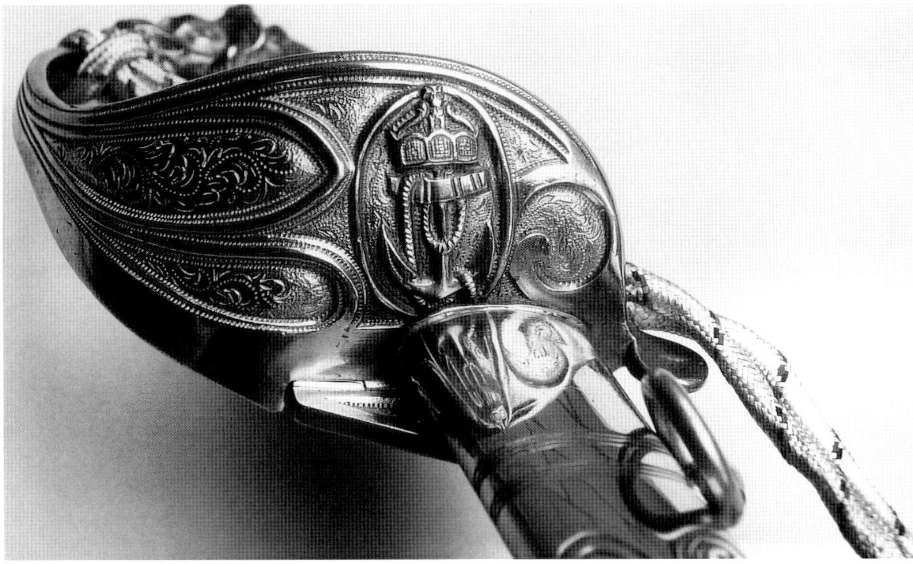

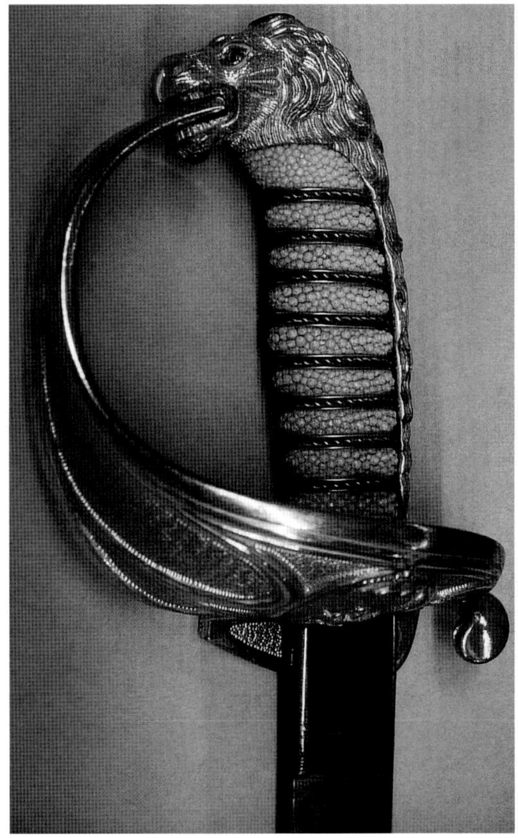

Above: W 68 Hilt of a naval saber of 1876, still much like the early type. But here the crown is almost as wide as the anchor stock and breaks out of the medallion at the top. The design of the crown indicates a later time, perhaps in the nineties. (PS)

Below left: W 69 Hilt of a naval saber of 1876. The Imperial crown is riveted on, thus not a cast component of the basket, and breaks out of the medallion at the top. The anchor with chain is a fixed (cast) component of the basket. The round tab to the right of the medallion is still flat as on the Prussian naval saber. (PS)

Above Right: W 70 Hilt of an Imperial naval saber. Instead of an ivory grip, this splendid weapon has soft fish skin. Made by W. Clauberg in Solingen. (DKS)

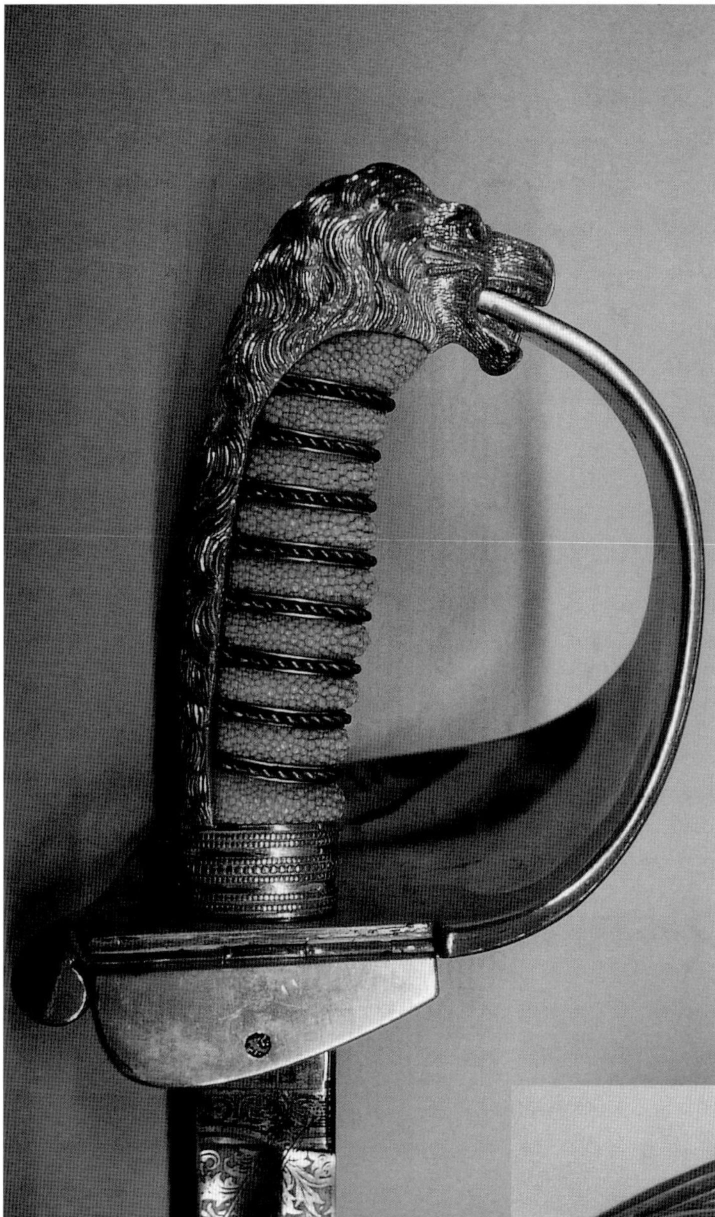

W 72 with especially finely engraved lion's head with eyes like a doll (in collectors' jargon: teddy-bear eyes). (DKS)

Above: Back of the saber in W 70. The finely worked grip ring is noteworthy. (DKS)

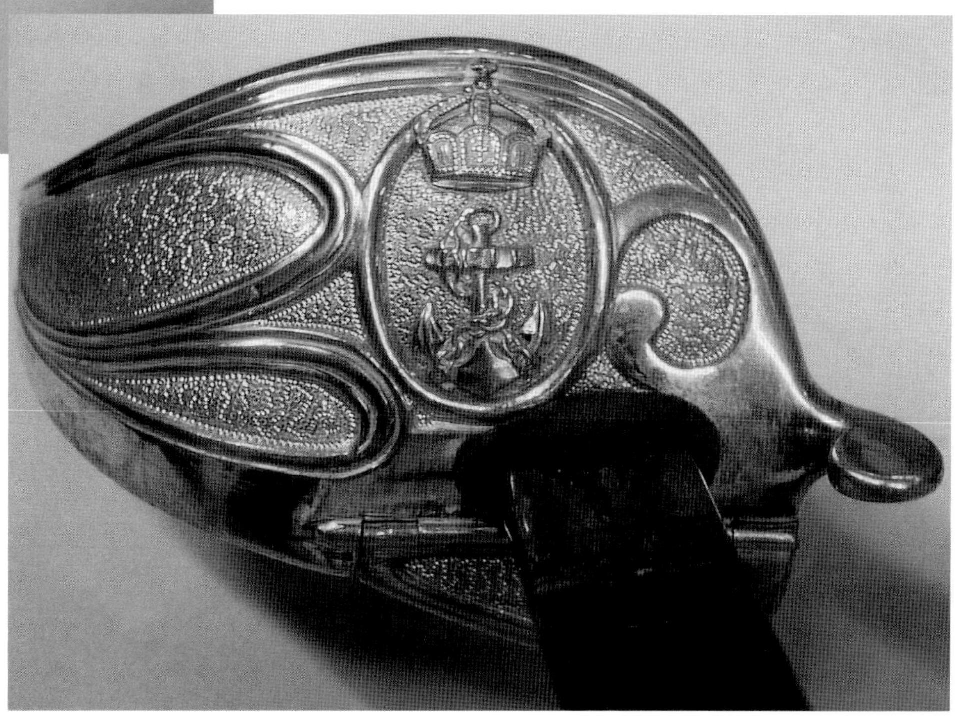

W 73 basket of the fancy saber. Here too, the added crown (as in W 69) breaks out of the medallion; the anchor with chain is a fixed component of the grip. (DKS)

Left: W 75 basket of a luxurious Imperial naval saber. The spaces between the decorative lines, which still correspond to the service model, are filled with tendrils and starfish. (PS)

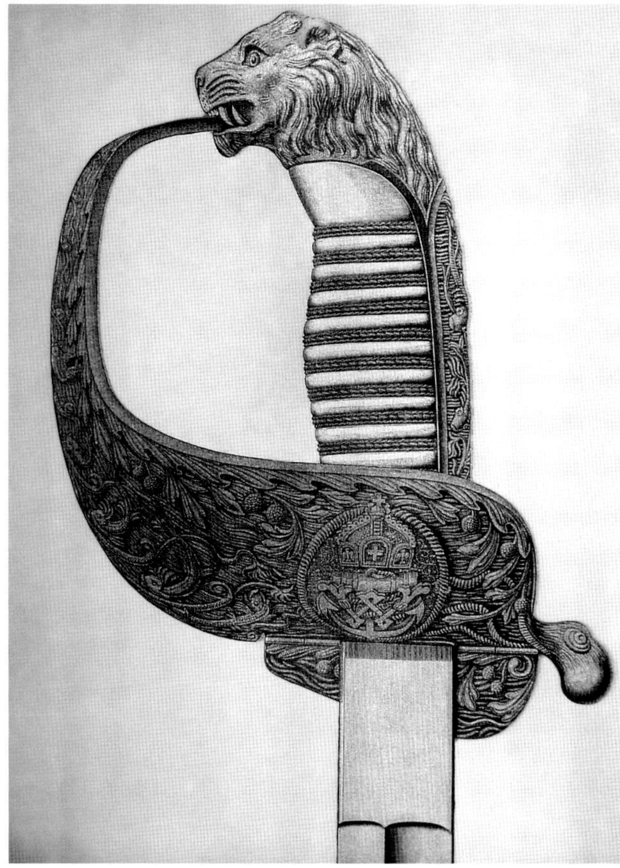

W 76 Imperial naval saber, from a pattern book (circa 1900) by Carl Eickhorn of Solingen. The luxuriously decorated weapon also has the initial "W" on the unclear anchor as a sign of respect for the German Kaiser. (DKS)

Upper left: W 74 basket of a naval saber of 1876 (private possession). The design of the hilt resembles the newer ordnance weapons. The crown is equipped with Pendilien. (PS)

W 77 Basket of a luxurious Imperial naval saber. The decorative lines are gone, and the hilt is decorated with a collection of reed leaves, sea roses and (imaginary) fish. (PS)

W 78 Hilt of a rare Imperial naval saber. (PS)

W 79 Basket of the weapon in W 78, see p. 69. The basic form of the basket still represents the prescribed type, but its formation is extraordinary. The rim of the basket is decorated with arabesques (seaweed?). Floral decorations occur on the bow as far as the motifs. On the left side (top view) of the basket is a cruiser before a rising sun. The usual anchor motif in the medallion is replaced by Kaiser Wilhelm II. (PS)

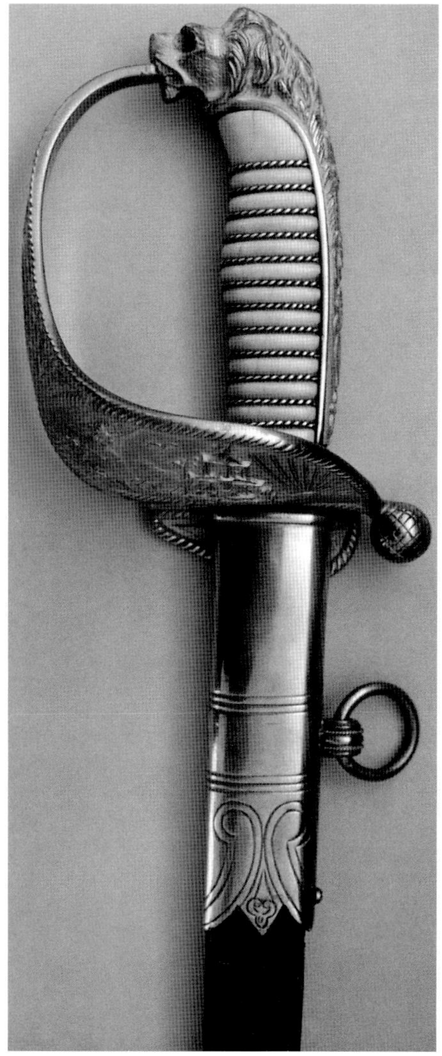

W 82 Hilt of the rare Imperial naval saber with feminine form (W 81). The end of the parrying bar is extraordinary, being formed like a globe. (PS)

W 80 Details of the basket of the weapon in W 78, see p. 69. On the right side of the basket, under the Imperial crown, is a diagonal anchor, its cable forming an oval around the figure of Kaiser Wilhelm II. Around the monarch's head is a quotation that he made in September 1898 at the opening of the new harbor in Stettin: "Our future lies on the water." (PS)

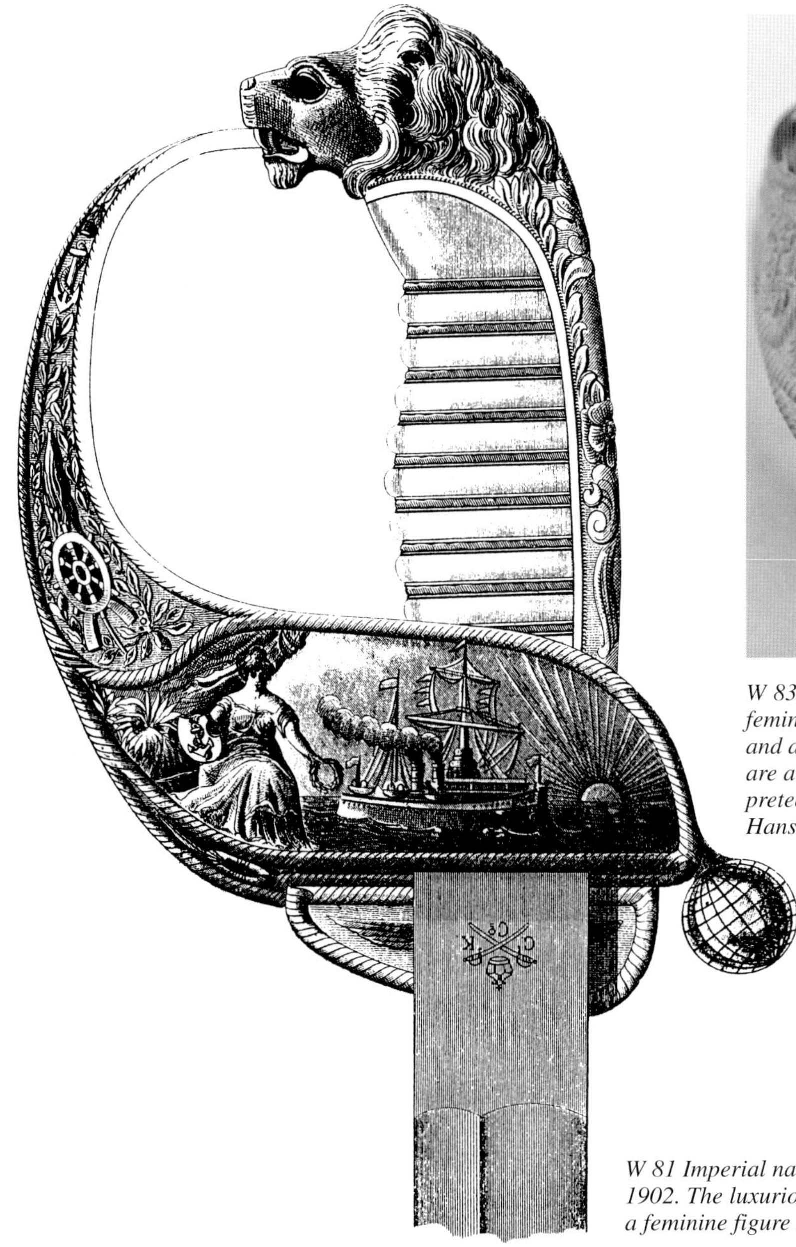

W 83 Basket of the weapon in W 82. To the left on the basket is a seated feminine figure, holding a shield with an unclear anchor in her right hand and a laurel wreath in her left. In the central and right parts of the basket are a cruiser on the high seas before a rising sun. The motif can be interpreted variously. The feminine figure could be Hammonia (symbol of the Hanseatic city of Hamburg) waiting for her "sons" who travel the world. The globe (Hamburg as "door to the world") supports this concept. On the other hand, the feminine figure could just as well be an allegory representing hope. This symbol appears here and there in maritime things. (PS)

W 81 Imperial naval saber, from a pattern book of the Carl Kaiser firm in Solingen in 1902. The luxuriously designed weapon likewise has extraordinary basket decor, with a feminine figure before a cruiser, this in turn before a rising sun. (DKS)

W 84 The bow of the weapon in W 82. The upper part bears an unclear anchor with an Imperial crown above it.

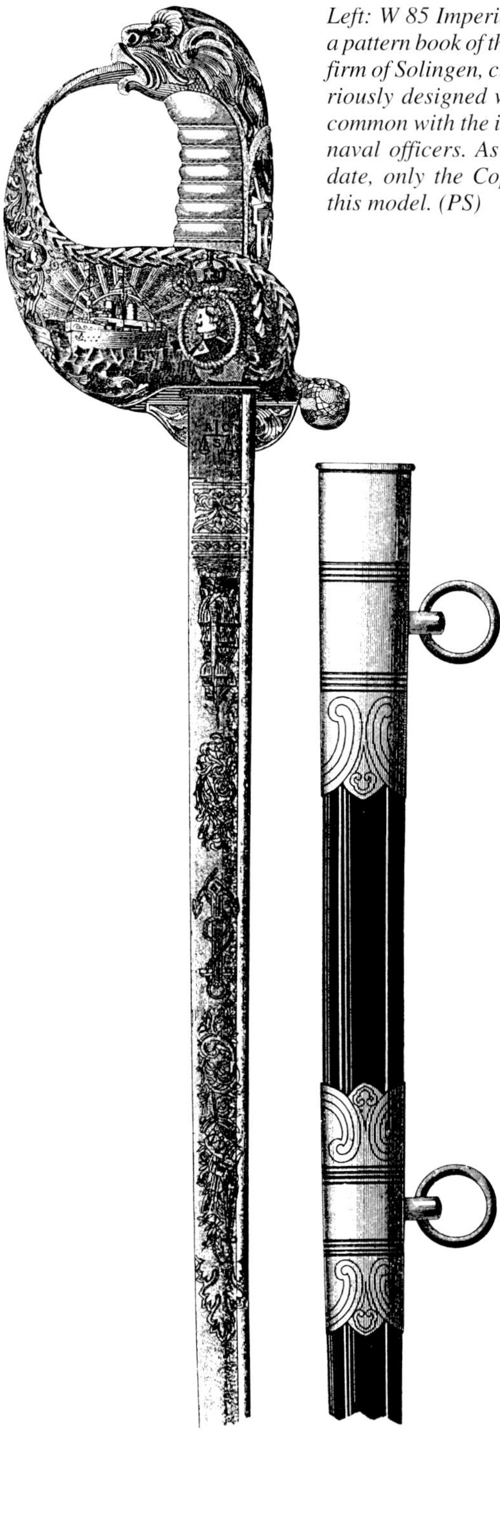

Left: W 85 Imperial naval saber from a pattern book of the Alexander Coppel firm of Solingen, circa 1910. The luxuriously designed weapon has little in common with the introduced saber for naval officers. As far as is known to date, only the Coppel firm produced this model. (PS)

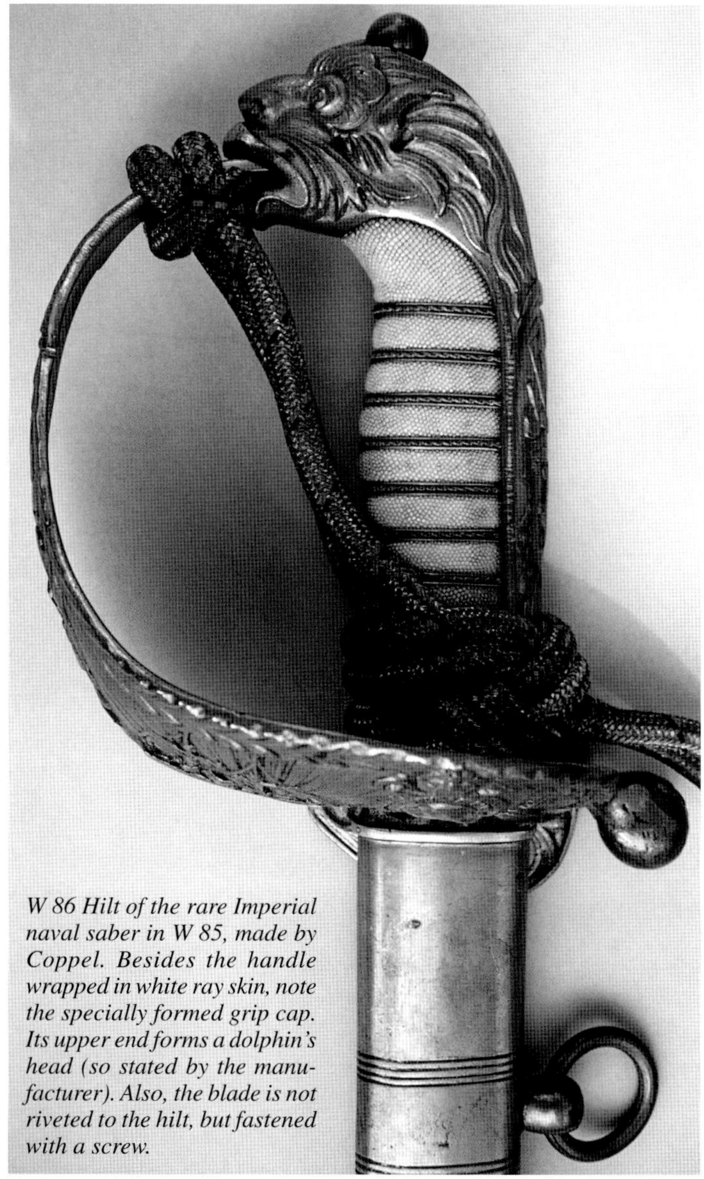

W 86 Hilt of the rare Imperial naval saber in W 85, made by Coppel. Besides the handle wrapped in white ray skin, note the specially formed grip cap. Its upper end forms a dolphin's head (so stated by the manufacturer). Also, the blade is not riveted to the hilt, but fastened with a screw.

Right: W 87 Close-up of the basket of the W 86 weapon. On the right part of the basket, under the Imperial crown, is a diagonal anchor, its cable forming an oval around the picture of Kaiser Wilhelm II. Over the monarch's head is the quotation that he spoke at the opening of the new harbor in Stettin in September 1898 "Our future lies on the water." (PS)

Above: W 88 Close-up of the dolphin's head on the W 86 weapon. Obviously the artist who designed the shape of the grip cap had never seen a dolphin. The model for the head was copyrighted by Alexander Coppel. (PS)

Left: W 89 Close-up of the back of the W 86 handle cap. In the center is a ship's wheel on a shield, which in turn rests on aquatic plants (reeds). (PS)

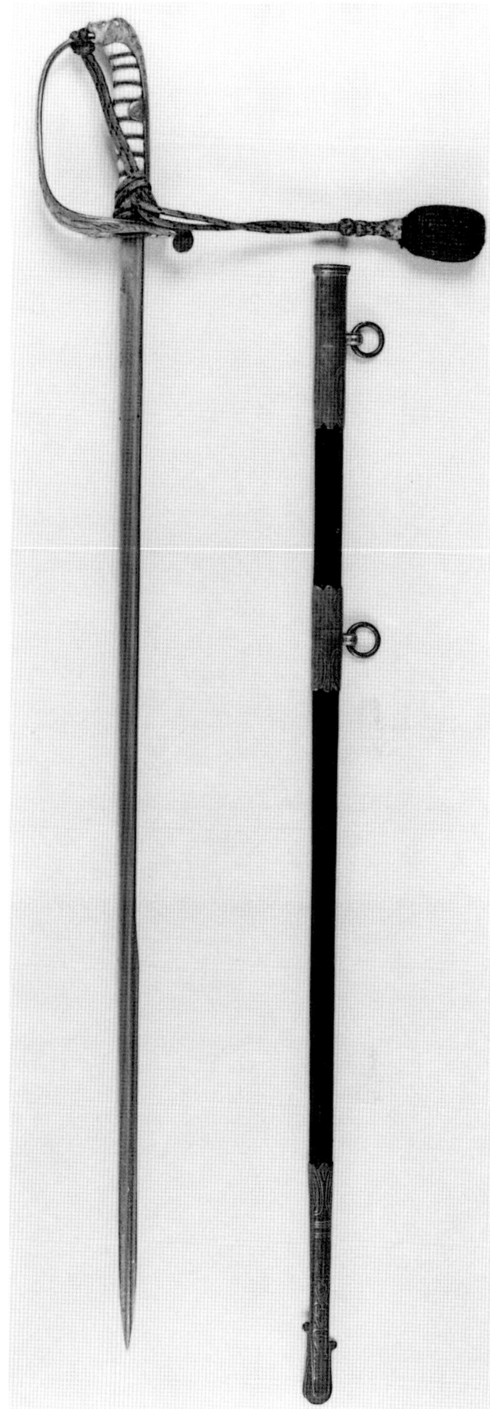

Right: W 90 Imperial naval saber. Note that the weapon does not have a curved blade, as sabers otherwise do, but a straight blade. It is a reinforced blade with a brace. (WGM: Inv. No. 105527)

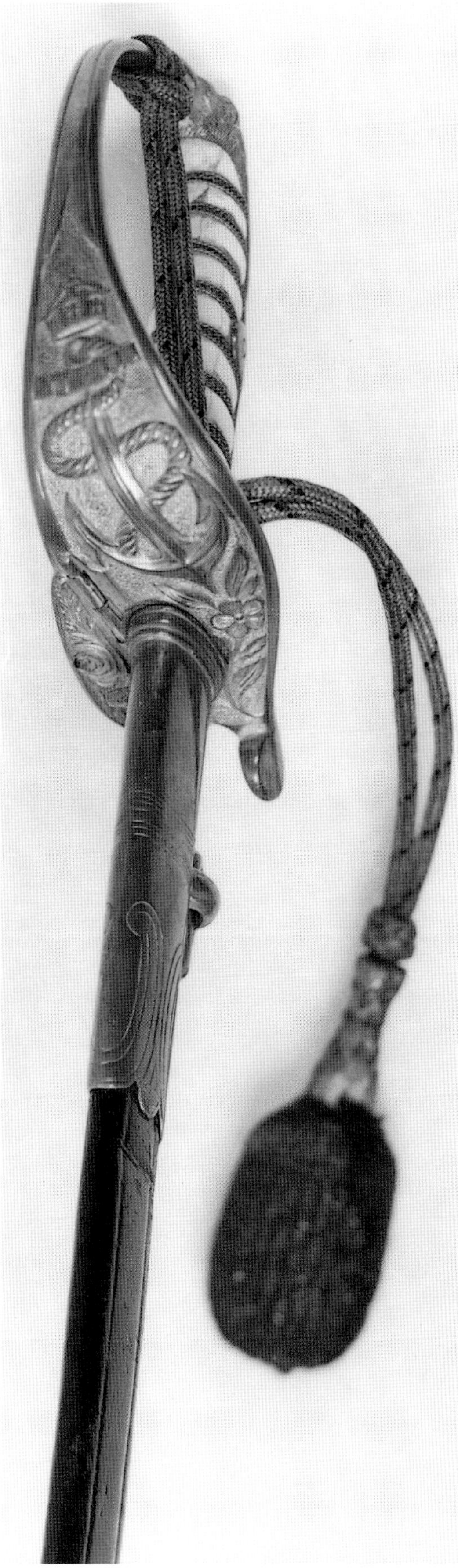

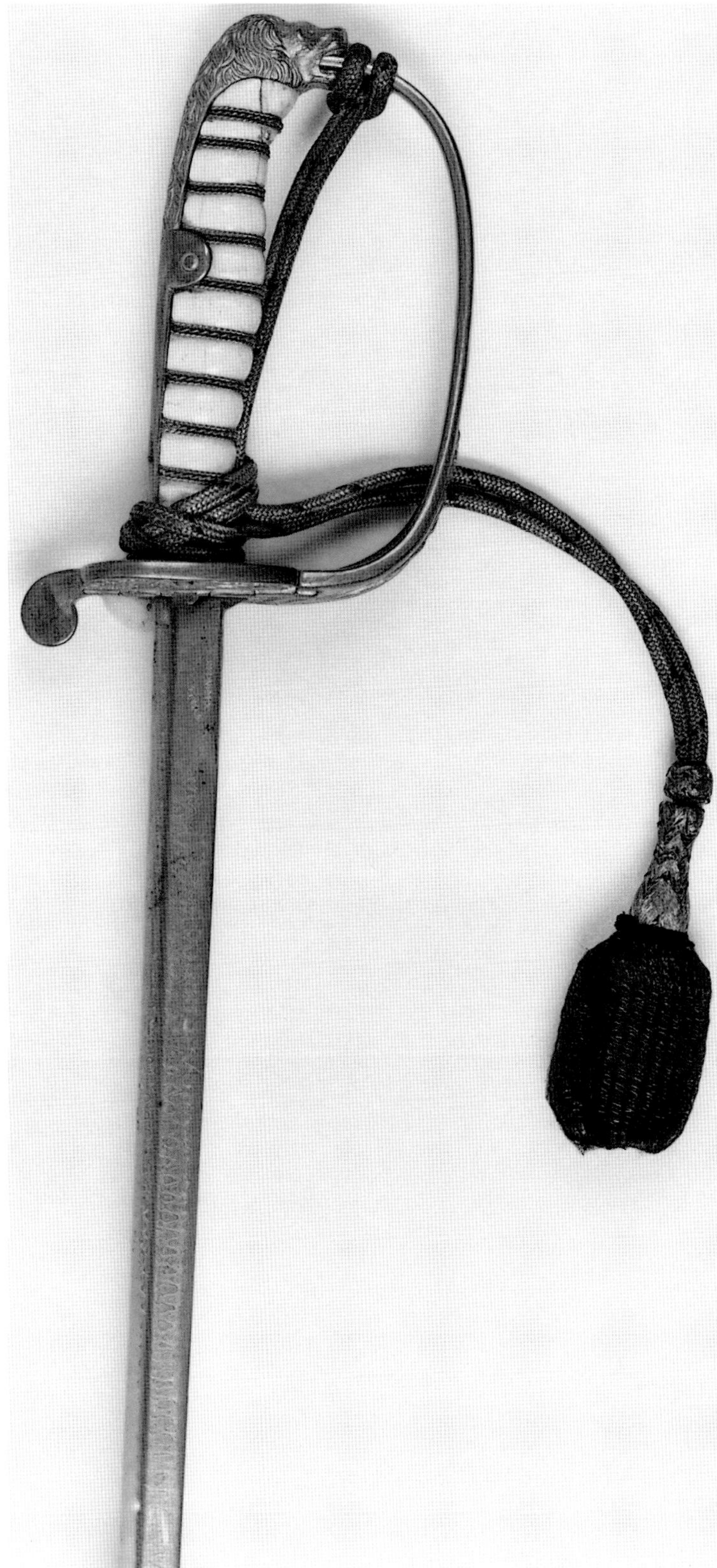

W 92 Quartside view of the hilt of the W 90 saber (p. 73). Here it is clear that the handle cap of the saber is fitted with cap flanges. This is unusual for an Imperial naval saber. (WGM: Inv. No. 105527)

Left: W 91 Hilt of the rare Imperial naval saber (W 90, p. 73). There is scarcely a basket. The bow turns into a guard with Imperial crown, unclear anchor, and floral decor. (WGM: Inv. No. 105527)

W 93 The former owner of the unusual weapon (W 92), Rear Admiral von Hintze. It is rare when one can portray the former owner with his weapon. (PS)

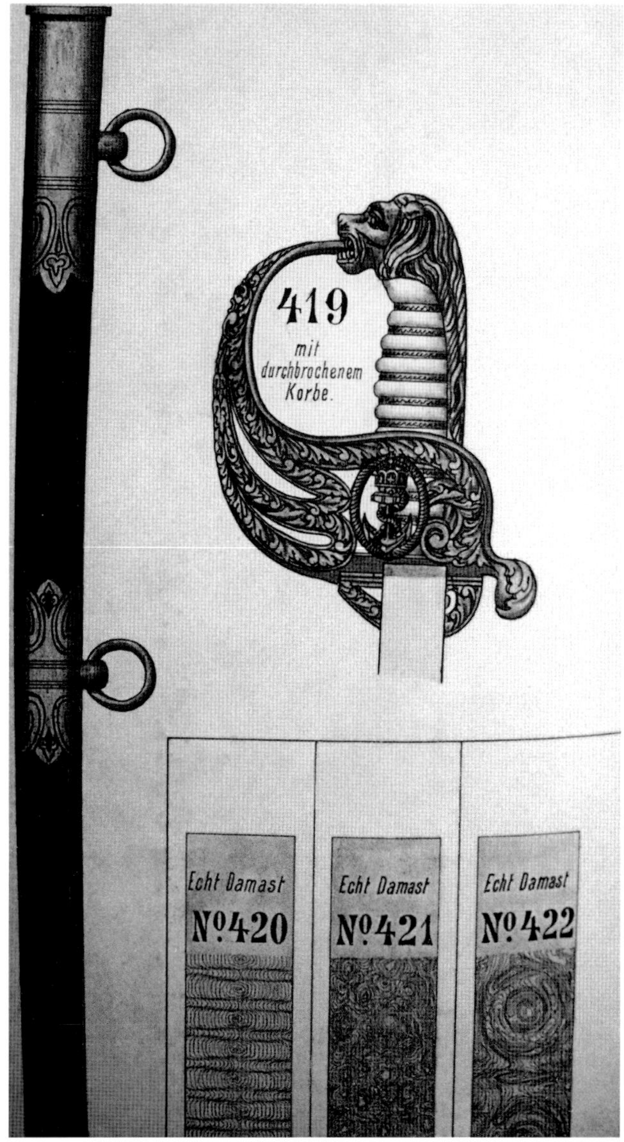

Right: W 94 A non-allowable variant of an Imperial naval saber, in a picture from the drawn and colored pattern book of Carl Eickhorn of Solingen (circa 1900). The pierced basket was not allowed for sea officers according to the introduction criteria of 1876. All the same, the Solingen sword industry delivered anything that pleased and—more importantly—was paid for. (DKS)

W 95: A forbidden pierced basket, as shown in W 94. This weapon was presented to Lieutenant Hedemann on his deaprture by his comrades of the II. Sea Battalion and Naval Field Battery. (MDG)

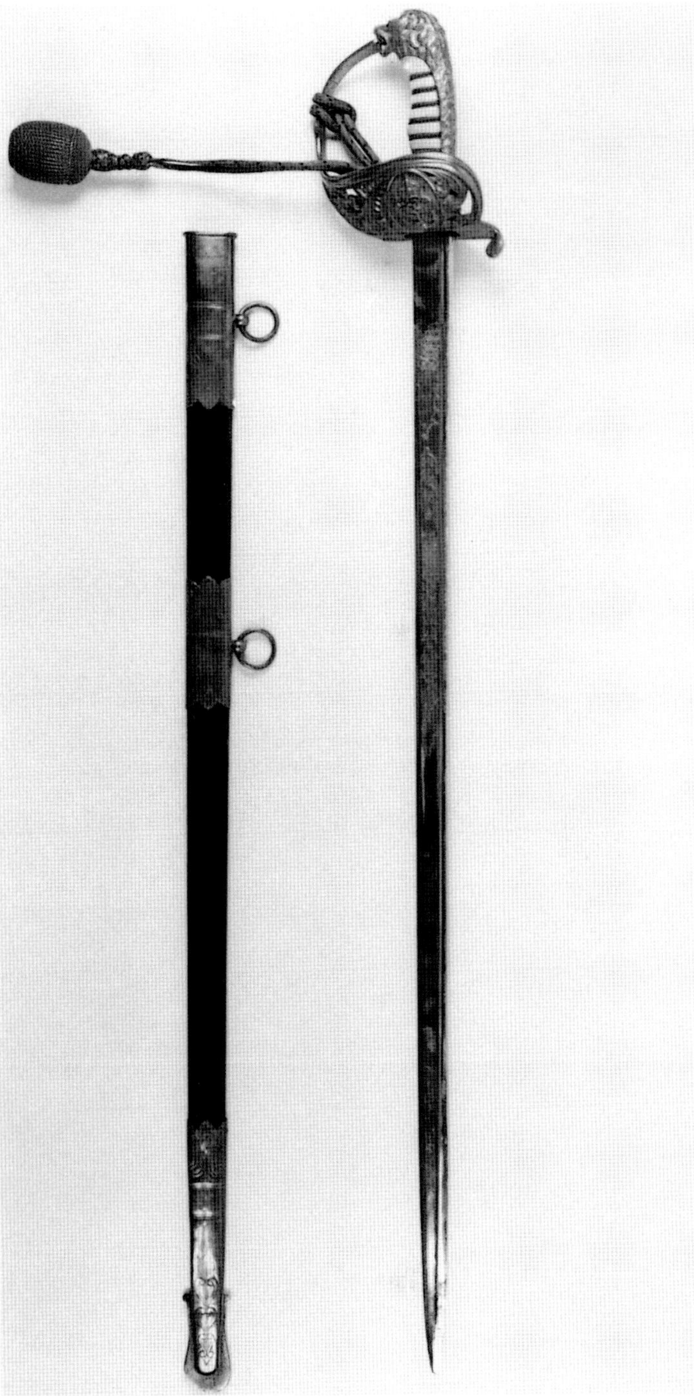

W 97 Imperial naval saber, not an allowed version, see the forbidden hilt in the next picture. Note that this weapon does not have a curved blade, like a customary saber, but a straight one. It has a reinforced back, ground in a three-sided shape. (WGM: Inv. No. 105525)

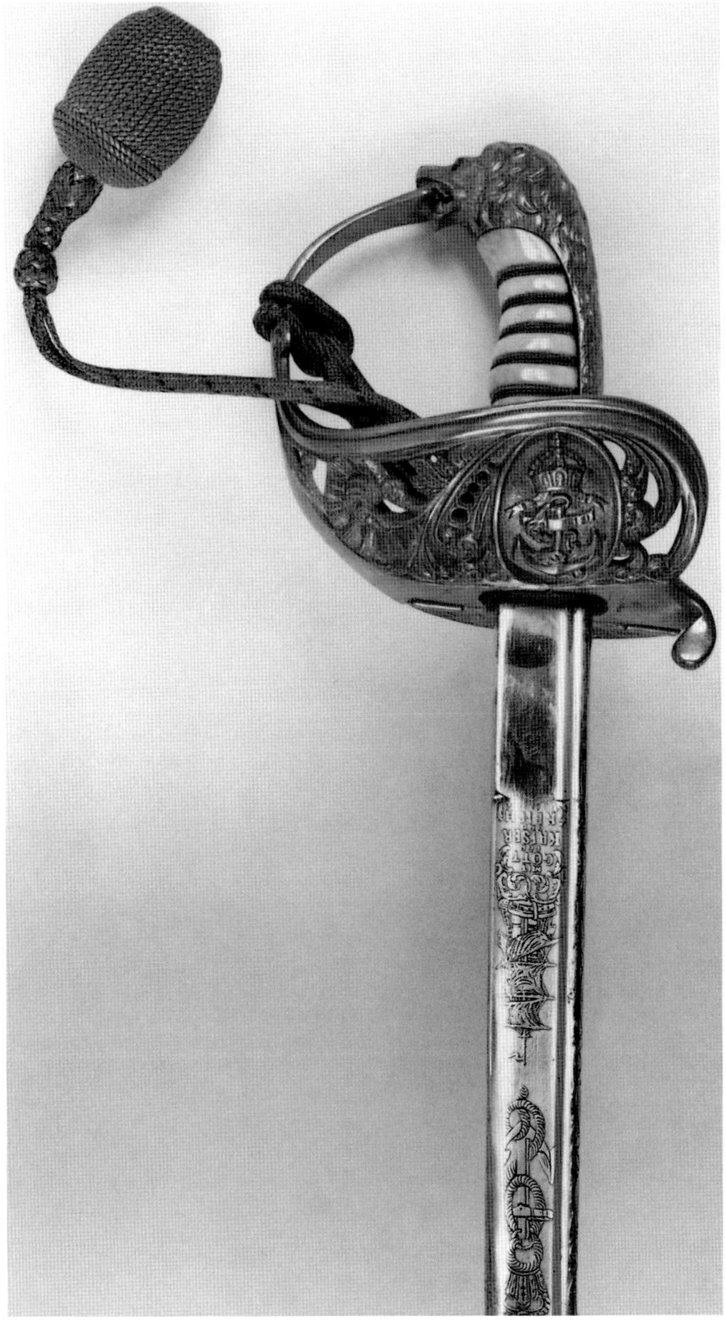

W 98 Hilt with pierced basket. The crown is the early type, but with Pendilien. The crown over the unclear anchor on the blade is also the early type. Between the dull edge and the sailing ship is the slogan stemming from the Prussian era: "With God for King and Country," now suited to Imperial times: "With God for Kaiser and Reich." (WGM: Inv. No. 105525)

W 96 One side of the blade inscription of the weapon in W 95 (see p. 75): The officer corps of the II. Sea Battalion and the Naval Field Battery—China 1900-1901. (MDG)

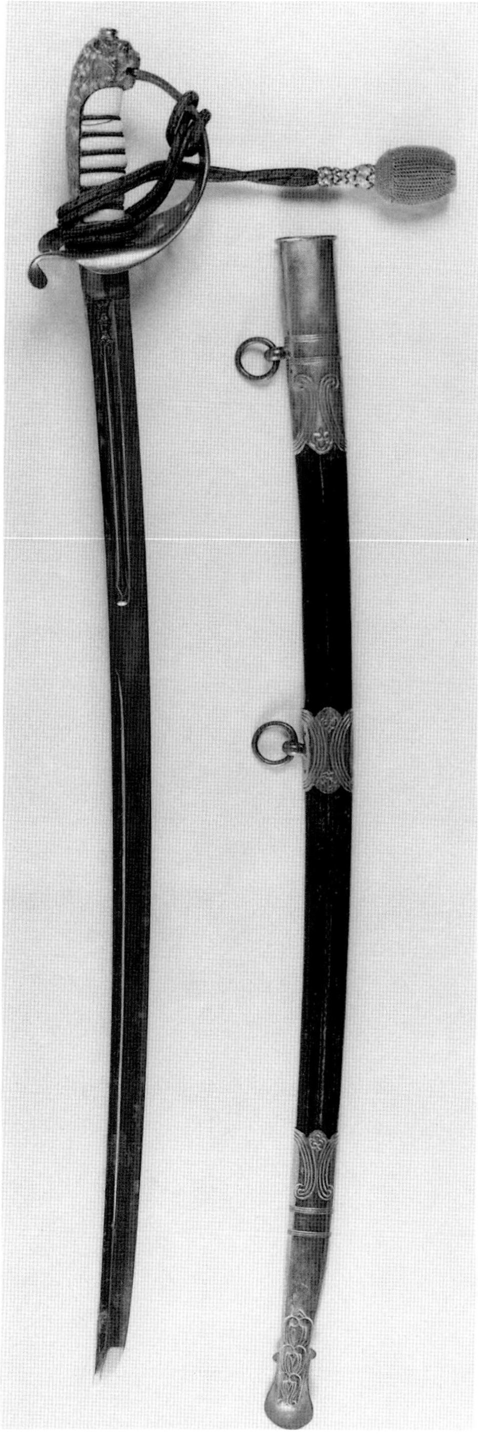

W 99 Imperial naval saber in non-allowed form. Note that this weapon's blade is not European but Japanese. It was given to its original owner, Rear Admiral von Hintze, on a visit to Japan. Japanese blades turn up here and there among naval and army weapons. Military men who saw service, for example, in China and returned home via Japan came to admire the blades there. See the next picture for the hilt. (WGM: Inv. No. 008461)

Right: W 100 Hilt of the saber in W 99, with pierced basket. The crown is still the earlier type. Note the large lion's head, and the guard that bends downward. (WGM: Inv. No. 008461)

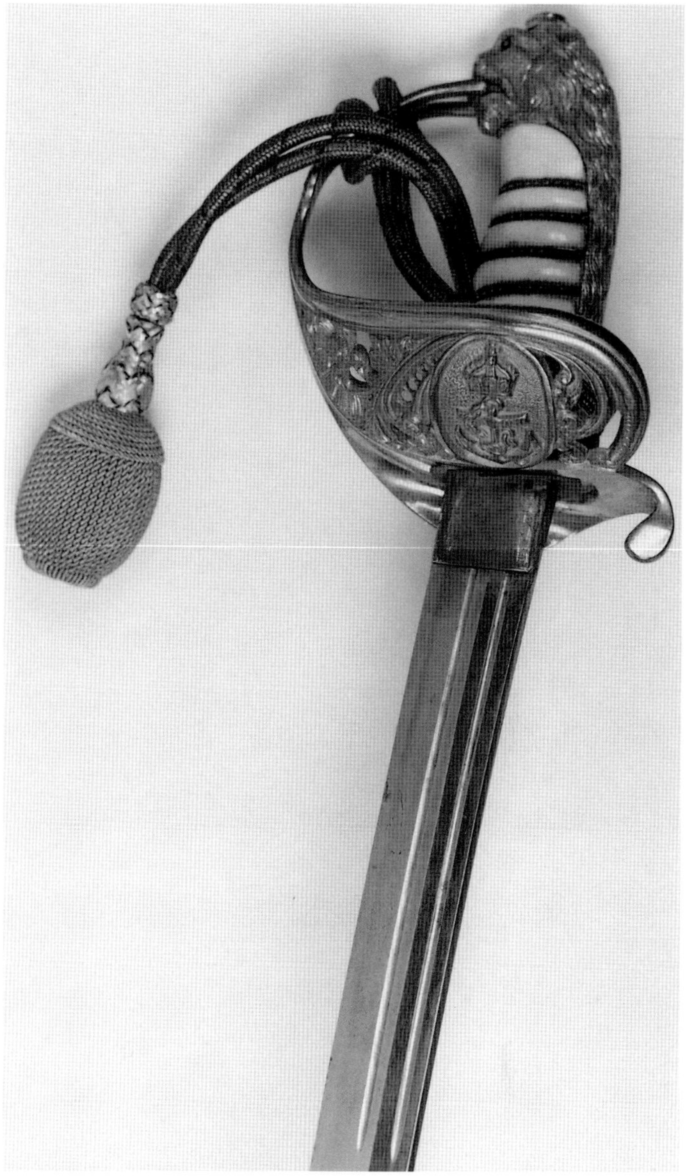

W 102 Basket of the saber in W 101 (see page 78). Note the strikingly large and somewhat three-dimensional anchor and cable with the early Imperial crown above. These motifs are wreathed on both sides by blooming laurel sprigs. (PS)

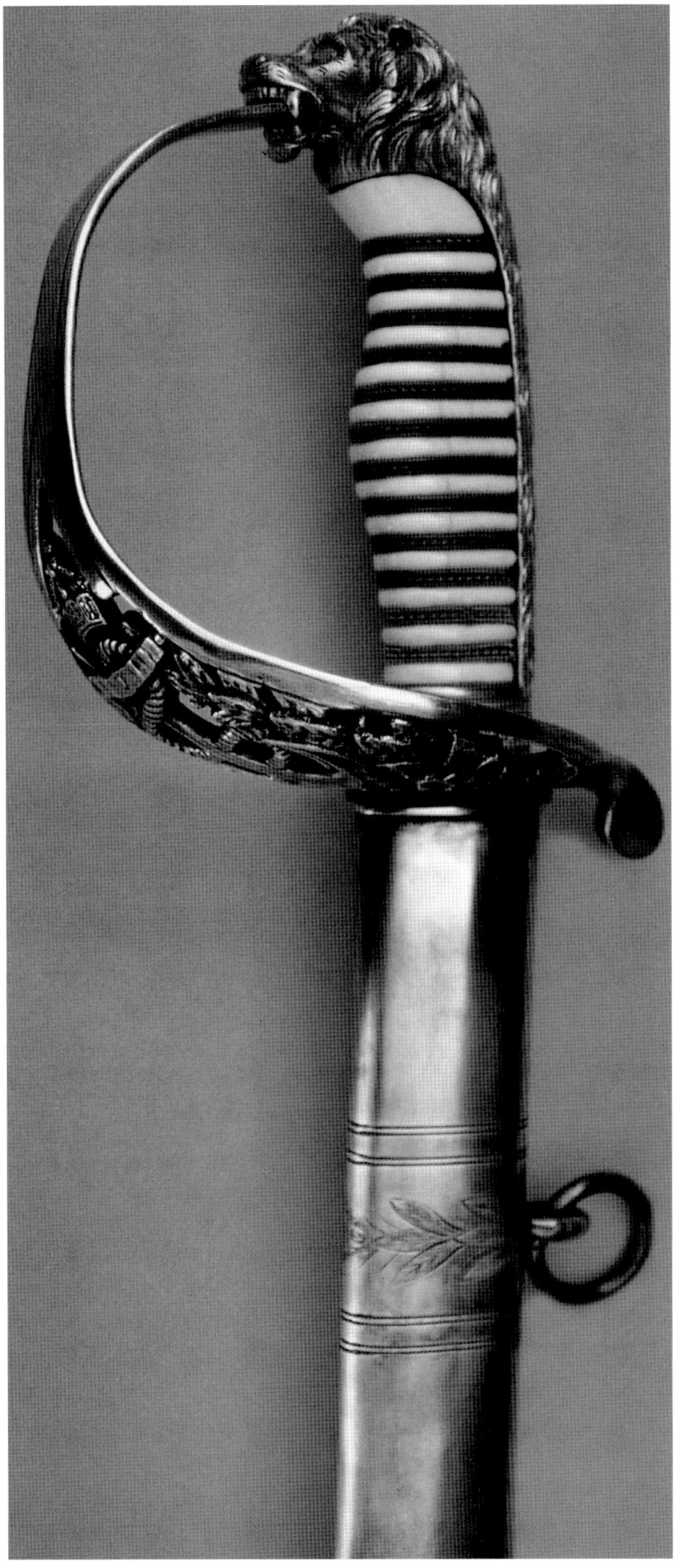

W 101 Imperial naval saber of a non-allowed type. Here too, the hilt is pierced. (PS)

W 104 The basket of the W 103 saber (see page 119, color section). The decorative lines correspond to the service model. The crown in the medallion is still the earlier type, but the formation of the unclear anchor suggests later times. (PS)

W 105 Imperial naval saber of a special design. The lion's mane is worked particularly well. (PS)

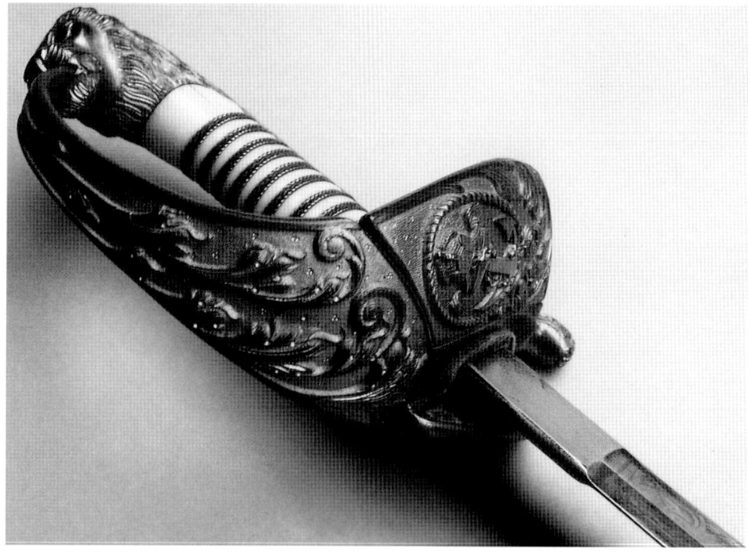

W 106 Basket of the W 105 saber. The bow is decorated with arabesques as far as the flap. The trapezoidal flap has a diagonal unclear anchor lying in an oval formed by the chain, which the anchor has instead of a cable.

W 107 The basket of an Imperial naval saber. Here too, the trapezoidal flap has a diagonal unclear anchor with a chain, surrounded by arabesques. (SPT)

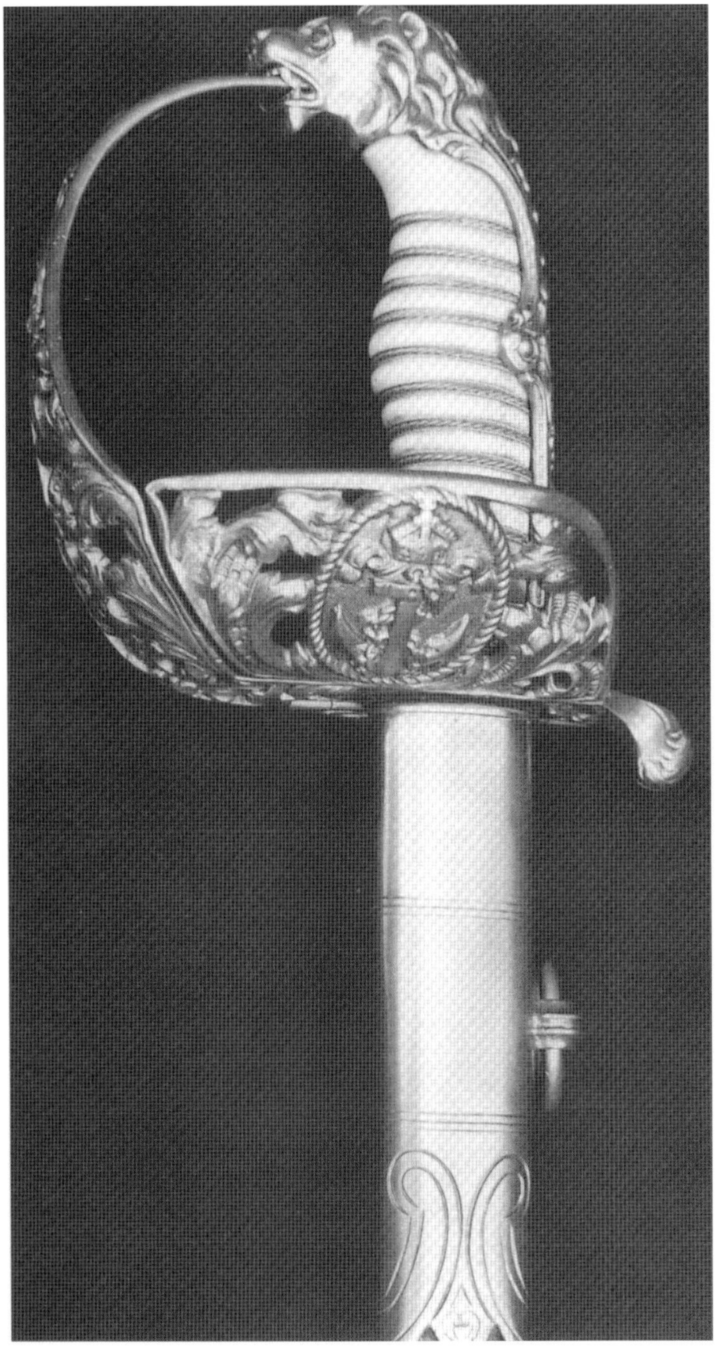

W 108 Imperial naval saber of a non-allowed type. The trapezoidal flap is pierced and has a standing unclear anchor in a medallion, and above it an Imperial crown with Pendilien (not a part of the hilt casting). Here too, arabesques decorate the hilt. (PS)

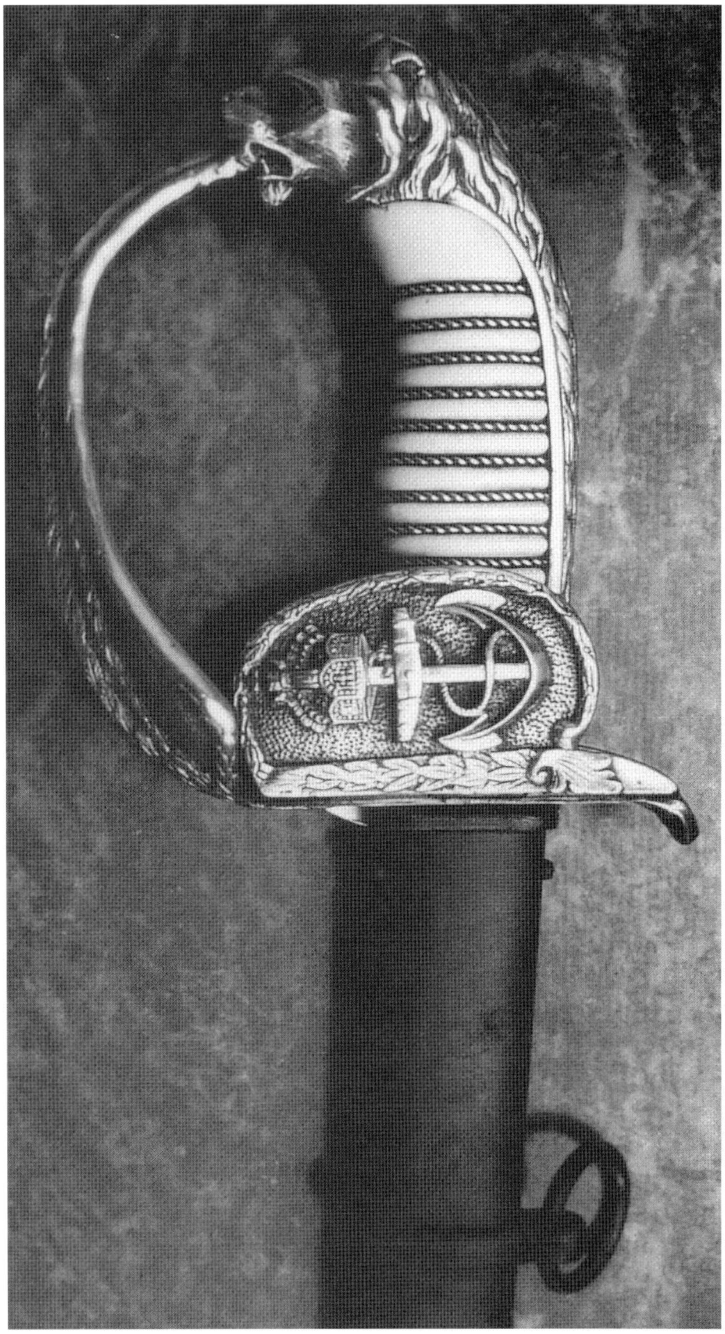

W 109 Imperial naval saber. The hilt has an oval flap. (PS)

Left: W 110 Imperial naval saber. This hilt also has an oval flap, which is richly worked. The unclear anchor (later form) is horizontal, with a crown above it. (PS)

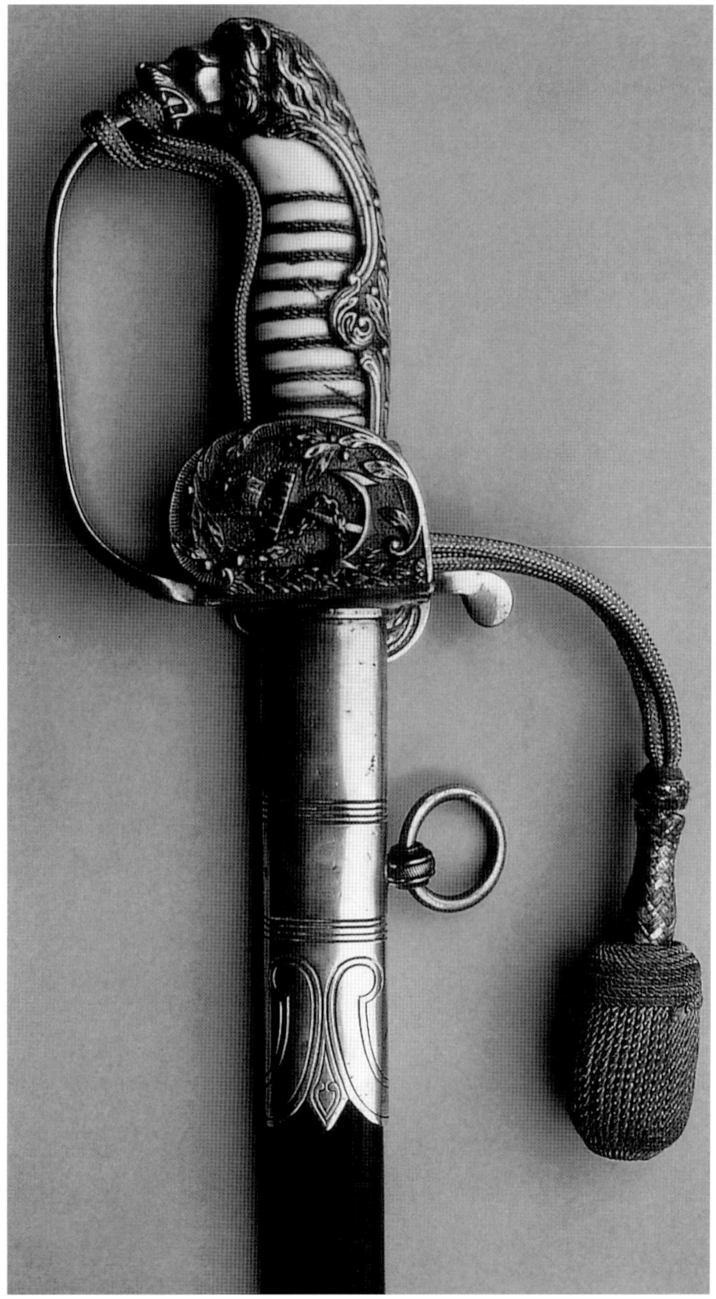

Above: W 111 Imperial naval saber. The anchor is diagonal on the flap.

W 112 Hilt of the W 111 naval saber (page 81). The diagonal unclear anchor is surrounded by laurel sprigs. The bow also has corresponding decoration. The portepee is the later type. (SPT)

Upper right: W 115 Bow of the W 114 weapon. Here too, the pierced oak-leaf decoration can be seen clearly. (PS)

Left: W 114 Imperial naval saber in non-allowed form. The hilt is pierced. The flap has a medallion, surrounded by oak leaves, and containing the vertical unclear anchor topped by a crown. (PS)

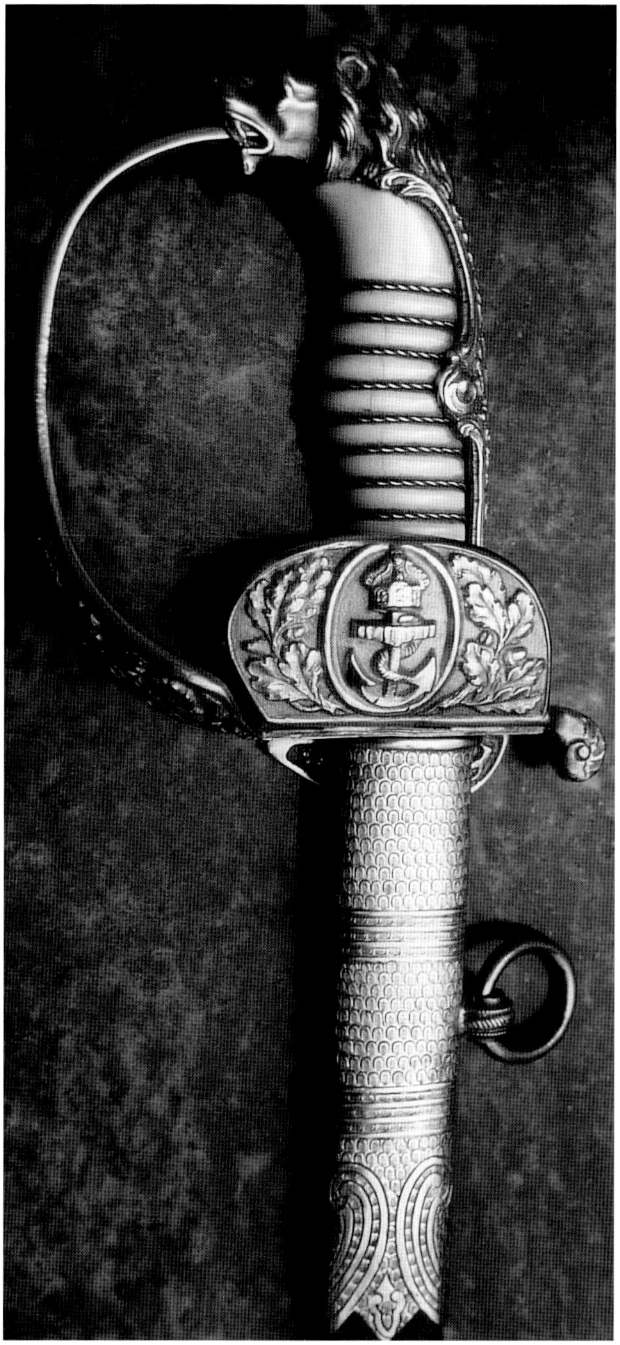

W 116 Imperial naval saber similar to W 114. The hilt, though, is not pierced. Note the especially finely decorated scabbard. (PS)

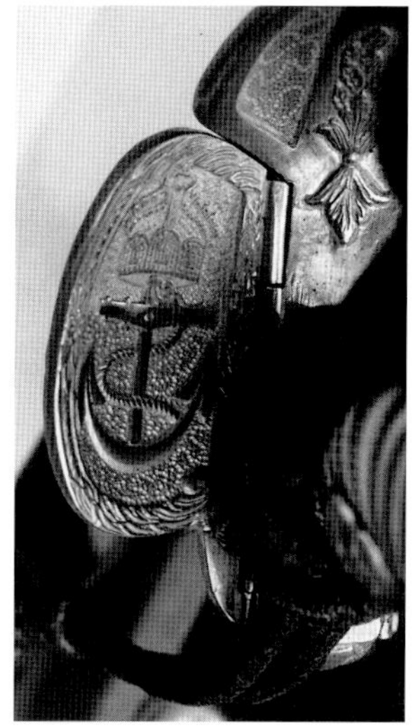

W 117 Flap of a luxurious Imperial naval saber. In rare cases, as here, the inside of the flap is also finely engraved. (PS)

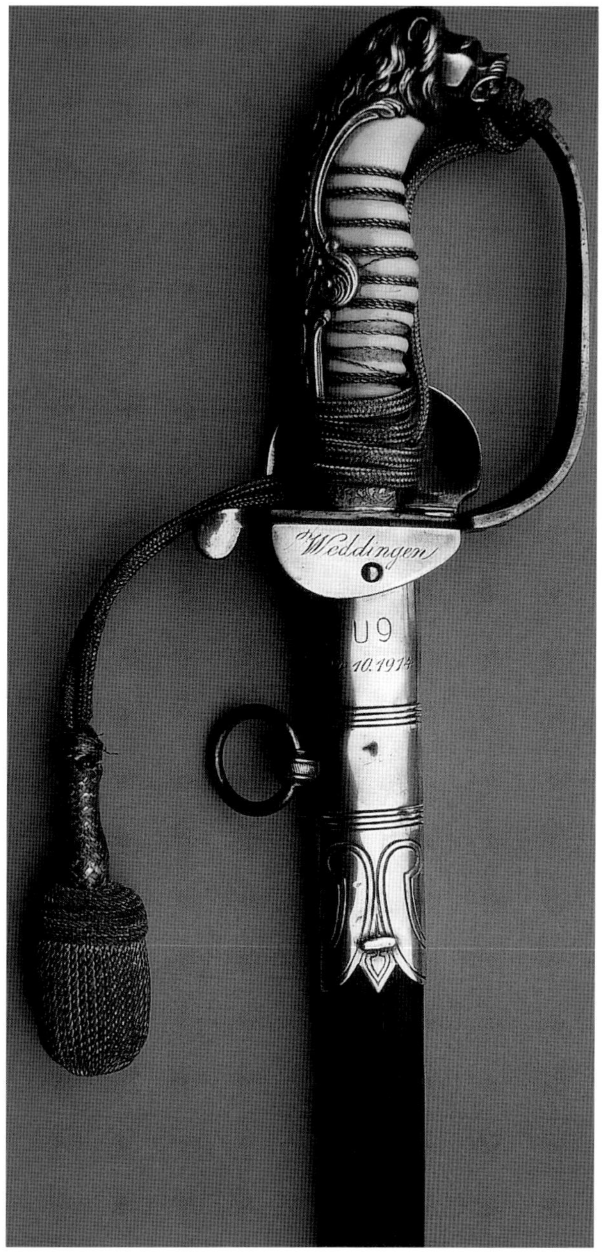

W 118 The quartside flap of an Imperial naval saber. The bore-hole serves to hold a spur soldered to the scabbard, to prevent unintentional parting of the weapon and scabbard. The flap was often used to bear the engraved name of the owner, here Weddingen. (SPT)

W 119 Quartside flap of an Imperial naval saber. The scabbard has an additional soldered-on carrying hook, so that the saber could also be worn vertically. (PS)

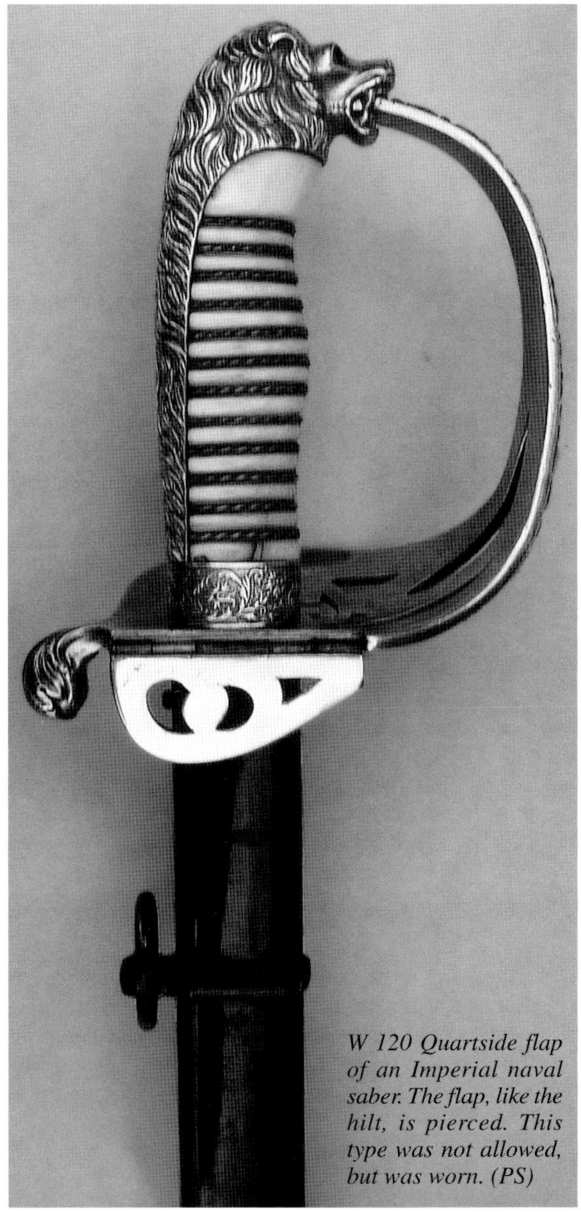

W 120 Quartside flap of an Imperial naval saber. The flap, like the hilt, is pierced. This type was not allowed, but was worn. (PS)

W 123 A luxurious Imperial naval saber. The stamp GES. GESCH.—here seen on the parrying bar—indicates that it is copyrighted. (SPT)

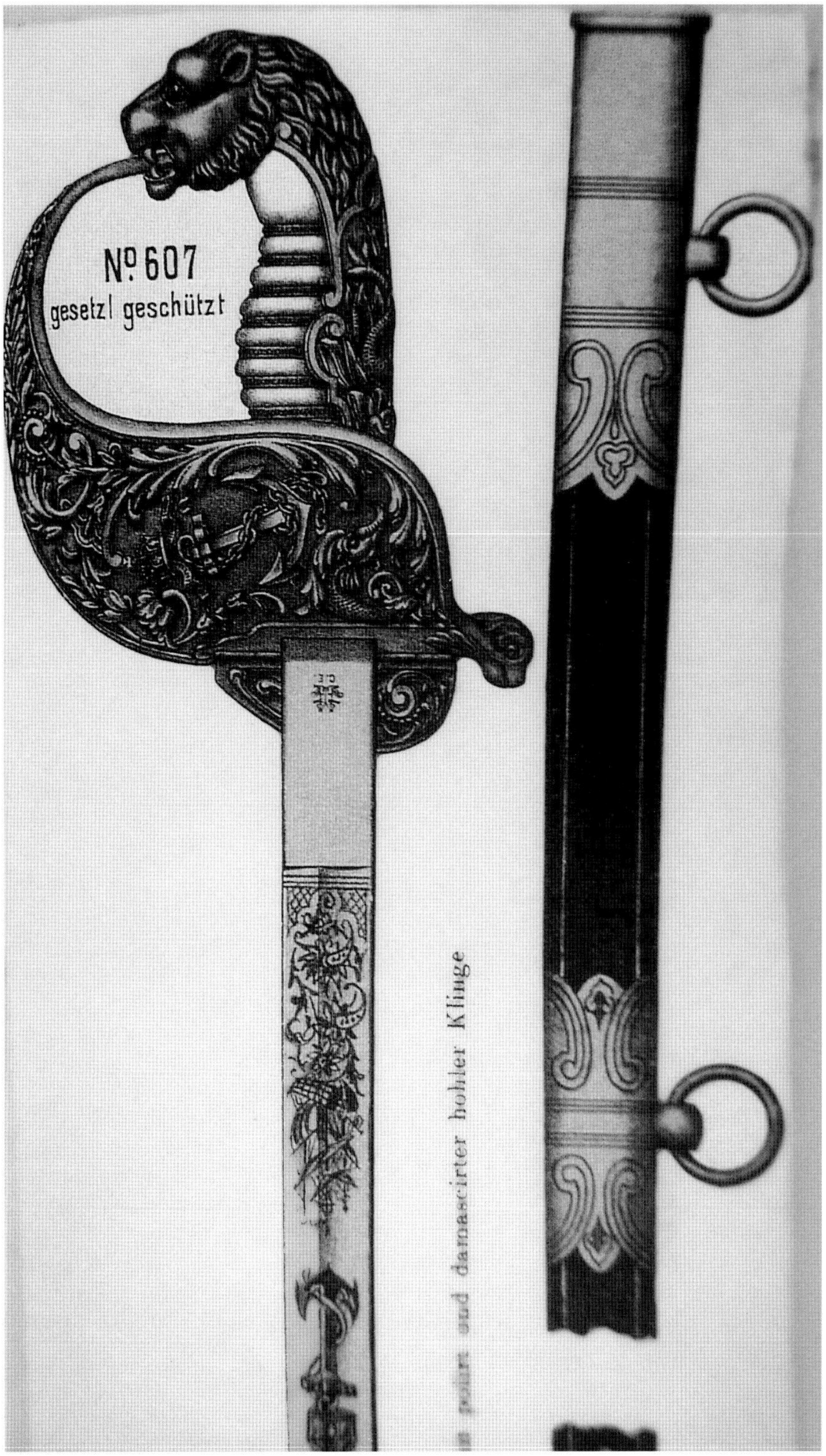

W 122 A luxurious Imperial naval saber, shown in the drawn and colored pattern book of Carl Eickhorn of Solingen (after the turn of the century). As can be seen, the firms registered and copyrighted designs developed by them. Then they made the pictures and weapons with the marking: gesetzlich geschu"tzt (legally protected). (DKS)

From left to right: W 124 Blade decoration of an early Imperial naval saber, with the early type crown over the unclear anchor. (PS) W 125 Blade decoration of a pre-1900 Imperial naval saber, with the later type of crown above the unclear anchor. (PS) W 126 Blade decoration of an Imperial naval saber from the beginning of the 20th century. (PS)

Left: W 127 Part of the blade decoration of a luxurious Imperial naval saber (same weapon as W 70, page 66): paddles behind unclear anchor.
Right: W 128 Part of the blade decoration of a luxurious Imperial naval saber (same weapon as W 70, page 66): compass on trophy with weapons. (DKS)

Above: W 132 Blade decoration of an Imperial naval saber, Damascus blade with dedication: Wodrig s./l. Kawelmacher (s./l. = seinem lieben: to his dea...). (PS)
Center: Blade decoration of an Imperial naval saber. Damascus blade with dedication: Der Freund dem Freunde (the friend to the friend). (PS)
Bottom: W 134 Blade decoration of an Imperial naval saber. On the back of the Damascus blade (W 133) is the dedication "Von Deinem Erich" (from your Erich). (PS)

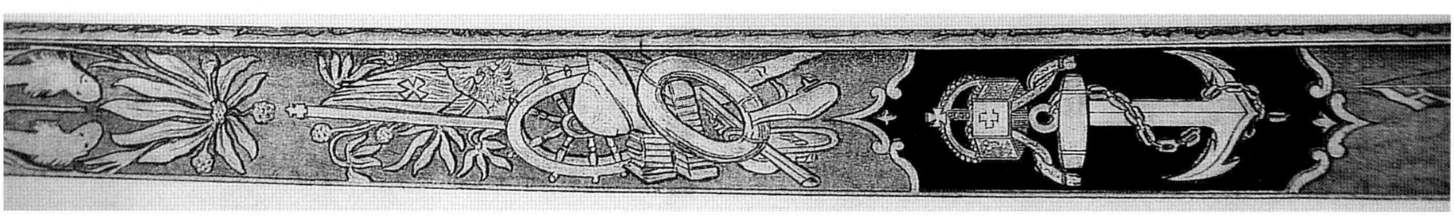

Above: W 130 Part of the blade decoration of an Imperial naval saber from a pattern book. An example of an inscription: "Das höchste Heil, das letzte liegt im Schwerte."
Below: W 129 Parts of the blade decoration of an Imperial naval saber from a pattern book of Carl Eickhorn. Along with the saying of Kaiser Wilhelm, "Unsere Zukunft liegt auf dem Wasser," the part of the engraving over the cartouche with the unclear anchor is interesting. (PS)

W 131 Blade decoration of an Imperial naval saber. On the right side of the Damascus blade (W 130) is the dedication" "Kilian, Nemitz, Mu"ller i. l. Roller (i.l. = ihrer lieben = to their dear...). (PS)

W 136 Blade decoration of an Imperial naval saber. On the back of the Damascus blade (W 135) are the owner's initials, with a crown above them, and below is the quality indication, "EISENHAUER" along with the maker's stamp "EWALD CLEFF SOLINGEN." (PS)

Center: W 137 Upper scabbard fitting of an Imperial naval saber. The mouth fitting extends downward. (PS)

W 135 Blade decoration of an Imperial naval saber. The blunt edge and the upper part of the blade (terzside) are finely etched and gilded. ECHT DAMAST tells the type of steel in the blade. (PS)

W 138 Upper scabbard fitting of an Imperial naval saber. The mouth fitting extends downward in steps. This type is often found in Army sabers too. (DKS)

W 139 Upper scabbard fitting of an Imperial naval saber, a finely worked version. (PS)

W 141 Central scabbard fittings of various Imperial naval sabers. Notice the differences in the decorative lines. (PS)

W 140 Upper scabbard fittings of various Imperial naval sabers. All are very elaborate designs. The scabbard on the left is from 1933-1945. (PS)

W 142 Central scabbard fitting of an Imperial naval saber. (DKS)

W 143 Central scabard fitting of an Imperial naval saber (W 139), with a finely conceived design. (PS)

W 144 Central scabbard fittings of various Imperial naval sabers (W 140). All are elaborately decorated. The scabbard on the left is from 1933-1945. (PS)

W 146 Lower scabbard fitting of an Imperial naval saber. Note the form of the drag plate. (DKS)

W 145 Lower scabbard fitting of an Imperial naval saber, the prescribed version. (PS)

W 148 Lower scabbard fittings of various Imperial naval sabers (W 140, page 88). All are very elaborately decorated. The scabbard on the left is from 1933-1945. (PS)

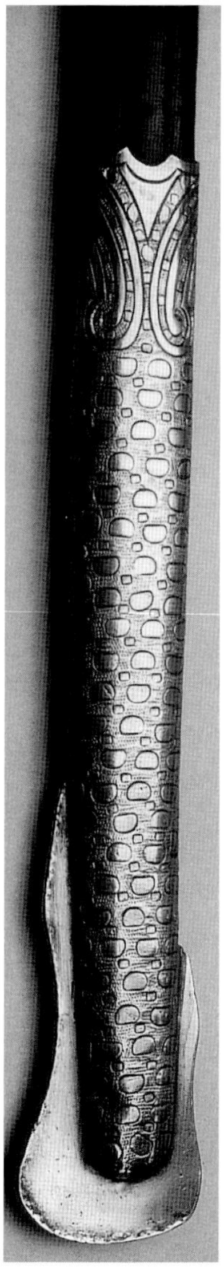

W 147 Lower scabbard fitting of an Imperial naval saber (W 139, page 88), with elaborate decoration. (PS)

W 149 Scabbard fittings of a luxurious Imperial naval saber, from a 1902 pattern book of the Carl Kaiser firm in Solingen. (DKS)

W 151 Special steel scabbard with decorative lines, as on the fittings of the leather scabbards, varying from the usual steel scabbards of naval sabers also in the soldered-on ring holder. (PS)

Left: W 150 Various types of fitting attachments. Left: The fitting was only glued on. Center: The fitting was also held on by wire. Right: This fitting was fastened with a special clamp. (PS)

To supply the officers with prescribed uniforms and equipment at reasonable prices, the Officer Uniform Commission of the Baltic Sea Naval Station had contracted with various uniform suppliers at the end of 1877. Certain quantities of sabers could also be obtained from these firms at set prices. Naturally, more had to be paid for luxury weapons. According to the contracts, between 27 and 54 Reichsmarks were to be paid for the "standard weapons." The price list of which sections for sidearms are reproduced here was made known to the Navy by the Chief of the Admiralty on January 3, 1878.[77]

On December 15, 1883, the *German Officer Union* was founded. This organization operated the *Warenhaus für Armee und Marine* (Store for Army and Navy) in Berlin, later also called the *Armeemarinehaus,* from which Army and Navy officers could order their uniforms and equipment. There were branch shops especially for the Navy in Kiel, Wilhelmshaven, and Flensburg. In the 1914 catalog naval officers' sabers were also offered. The prices at that time can be seen in the following overview.[78]

Bezeichnung.	Berger, Collani & Comp. I. Qualität M.		II. Qualität M.		Bemerkungen.	Schrand & Holstein. I. Qualität M.		II. Qualität M.		Bemerkungen.
Säbel für Offiziere 2c.	48,50	bis	27,50			45	bis	30		
Säbelkoppel do.	14	50	13	50		14	—	12	—	
Portepee do.	4	50	4	—		5	—	4	50	
Schärpen do.	34	—	27	50	inkl. Futteral von Pappe	32	—	30	—	inkl. Futteral von Pappe.
Säbel für Kadetten	13	—	—	—		10	50	—	—	
Säbelkoppel do.	5	75	—	—		7	50	—	—	
Portepee do.	4	—	3	50		4	—	—	—	

Bartsch & Bartling. I. Qualität M.		II. Qualität M.		Bemerkungen.	Lucas. I. Qualität M.		II. Qualität M.		Bemerkungen.	Geißler & Hast. I. Qualität M.		II. Qualität M.		Bemerkungen.
48	bis	27	—		54	bis	36			—	—	—	—	
16	—	14	—		17	—	16	—		—	—	—	—	
5	50	4	25		5	50	—	—		4	—	—	—	
31	50	28	50	inkl. Futteral v. Pappe.	33	50	—	—	inkl. Futteral v. Pappe.	27	75	—	—	inkl. Futteral v. Pappe.
14	50	12	—		14	50	—	—		—	—	—	—	
7	—	4	—		6	50	—	—		—	—	—	—	
4	30	4	—		4	50	—	—		2	75	—	—	

29 Two price lists from the naval regulation sheet.

Seitengewehre

a) Marine-Offizier-Säbel

Nr. 12268g	Vorschriftsmässig mit damaszierter Klinge		M 24,50
Nr. 12269	mit vergoldetem Beschlag u. Elfenbeingriff	nach Form II	M 40,50
Nr. 12270		" VI	M 42,50
Nr. 12271		" VIII (fein ziseliert)	M 45,—
Nr. 12272		" IX " "	M 45,—
Nr. 12273		" XI	M 45,—
Nr. 12274		" XIV (Doppelscharnier)	M 50,—
Nr. 12275		" XV " leicht)	M 50,—
Nr. 12276		" XVI	M 45,—
Nr. 12277		" XVII (leicht)	M 45,—
Nr. 12278		" XIX	M 45,—
Nr. 12279		" XXIII (leicht)	M 45,—
Nr. 12280		" XXIV	M 42,50
Nr. 12281	Mit damaszierter Damaststahlklinge . . . . . . mehr		M 8,75
Nr. 12282	" vergoldeter " . . . . "		M 7,—

Alle Säbel sind mit englischer Steckenrück - Eisenhauer - Klinge und mit Augen versehen, ausgenommen Nr. 12268g — Knochengriffe im Preise 6 M billiger. — Feuervergoldung erhöht den Preis um 12 M für das Stück und für **fein ziselierte** Monturen um 15 M das Stück.

Nr. 12283	Lederscheide, lose, mit vergoldetem ziselierten Beschlag	M 12,—
Nr. 12284	mit grauer damaszierter Damaststahlklinge . . mehr	M 8,—
Nr. 12285	" vergoldeter " . . . . "	M 6,—
Nr. 12290	**Säbelbezug** " "	M 2,—
Nr. 12291	Säbelfutteral aus braunem, imprägniertem Segeltuch, mit Ledereinfassung, Rindsledergriff und Riemen . .	M 4,—

Säbelkoppel

(Bei Bestellung ist stets die Leibweite anzugeben)

Nr. 12292	Oberkoppel - Leibgurt mit seidenem Band und vergoldeten Beschlägen	M 9,—
Nr. 12293	Geflochtene Trage- u. Schleppriemen für Säbelkoppel mit vergoldeter Kette und doppelten Karabinerhaken	M 9,50
Nr. 12295	Unterkoppel mit Leibgurt und geflochtenen Trage- u. Schleppriemen und vergoldeter Kette und doppelten Karabinerhaken	M 12,50
Nr. 12298	**Schnüre** mit Karabinerhaken, lose . . . . . . Paar	M 8,50

Portepees

Nr. 12299	Argentan III	M 2,75
Nr. 12300	Silberplattiert, mit echter Schnur, II	M 4,25
Nr. 12301	" " " " Eichel u. Schieber, I	M 4,75
Nr. 12302	Ganz echt Silber	M 8,—
Nr. 12305	Goldplattiert mit plattierter Schnur	M 4,50
Nr. 12306	" " besser plattierter Schnur	M 5,50
Nr. 12307	" " echter Schnur	M 6,50

30 Price list from the Warenhaus für Armee und Marine.

Sabers for Mounted Officers of the Naval Infantry and Field Battery

The Steel Infantry Basket Saber

As already mentioned, the officer corps of the Sea Battalion was completed with Army men. Army officers transferred to the Navy and left it after a specified period of time to return to the Army, as long as they did not leave military service altogether.

37 Major Baron von Liliencron, Commander of the III. Sea Battalion, with his adjutant. MSM Archives

31 Prince Heinrich of Prussia, second from left, and Sea Captain Meyer-Waldeck, Governor of Kiauchau, on horseback. Both wear naval officers' sabers. MSM Archives.

38 Naval Senior Assistant Doctor Riecke with a naval officer's saber and a white tropical uniform. MSM Archives

39 A sergeant of the naval artillery (pre-1894). SPT Archives

42 General von Soden, former Commander of the Embassy Guard in Peking, with his naval saber. Note the army officer's portepee on the weapon. MSM Archives

In many ways they held a position between Army and Navy, including uniforms and weapons. For example, the officers wore the uniform of the guards, but with white collars; on the other hand, they were to wear the naval officers' saber since 1862. As in the Army, staff officers and adjutants of the naval infantry were regularly mounted, and the naval saber was also worn when on horseback. Later, as of 1888 (Order of October 23) the company chiefs were also "made mounted."[79]

40 Captain Count von Sparr in the uniform of the German East African Guard troops, carrying his naval saber in a steel scabbard. Sparr was formerly with the naval infantry in Wilhelmshaven. Rcakelmann Archives.

In a handwritten note of January 22, 1884, Chief of the Admiralty von Caprivi declared:

His Majesty the Kaiser has deigned to state in an immediate message today that the mounted officers of the Sea Battalion may wear the steel infantry basket saber in and out of service.

On the same day, the highest naval command officials, the Imperial Command of the North Sea and Baltic Sea Naval Stations, were informed of the new regulations. Through this directive the mounted officers of the Sea Battalion were allowed to wear the type of saber that had already been introduced into the Army for the corresponding officers in 1849.[80] A few months later, on March 29, 1884, the Ministry of War informed the Imperial Admiralty, to complete the Kaiser's decision:

As for the determination of a certain model for the saber to be worn by officers in the field, His Majesty the Kaiser and King has deigned to disregard it, so that the officers will be spared special purchasing costs, but above all so that the opportunity remains open for the officers to be able to continue to wear the saber worn by either himself or an ancestor in earlier campaigns. For peacetime conditions, though, the presently valid norm shall be retained.

Whether sabers were worn frequently in the Sea Battalion must be doubted, since the personnel who had the right to wear one—and at first there was only one sea battalion—could be counted on one hand. On the other hand, the appropriate officers probably owned this weapon, since they were known to come from the Army.

The model of the steel infantry basket saber, also referred to as the Infantry Officer's Saber M 1849, is defined neither in the form of an undoubted historical description nor as a stamped test model, nor through an authentic picture.[81] On the other hand, in specialist circles a model has been identified which fulfills all the known conditions, and of which several examples exist, undoubtedly Prussian on the basis of their blade etchings. This model shall be described.

41 Group photo of members of the III. Sea Battalion in China. The officer in the center of the photo carries his naval saber in a dark steel scabbard, while two portepee lower officers to the left wear naval sabers in leather scabbards. Rackelmann Archives

Blade: The weapon has a reinforced blade with a brace. The form of the blade with its decorations varies. There are several examples that bear the letters FWR or WR[82] between floral decoration and military trophies, plus a Prussian eagle on the back. Examples also exist with blades without any decoration.

Hilt: Three terzside bows and one main bow form the hilt, along with the gently curved guard, which is made of steel. Under the grip ring, which has several decorative grooves and holds the handle and is covered with fish skin, there rises a white leather finger eyelet bordered in black. The handle has seven or eight windings, mostly of thin corded silver wire. The handle cap is decorated at its top end with six lance-shaped grooves.

Scabbard: The plain steel scabbard is carried by two rings.

This weapon was, moreover, also to be worn by the officers of the troops of Reich Commissioner *Wissmann,* according to the *Directions for the conducting of service in the stations of the guard troops for East Africa.* The appropriate order is dated November 4, 1890.

In the *Uniform Regulations for Sea Officers, Officers of the Naval Infantry,...* of 1893 the saber is still mentioned.[83] During the course of the following years, this model was dropped.

The Infantry Officer Sword 89

In 1889 Kaiser Wilhelm II—by order on March 22[84]—approved the introduction of a new sidearm for the officers of the foot troops, who had worn various sidearms before. This measure also had an indirect effect on the officers of the Sea Battalion.[85] The previously allowed steel basket saber was no longer to be purchased.[86] In order to go on benefiting from the advantages of a sidearm in a steel scabbard on horseback, the mounted officers of the Naval Infantry also had to take on the new sword.

W 152 Hilt of a steel infantry basket saber. This model is designated in present-day specialist literature as the Prussian Infantry Officer's Saber M 1849. (PS)

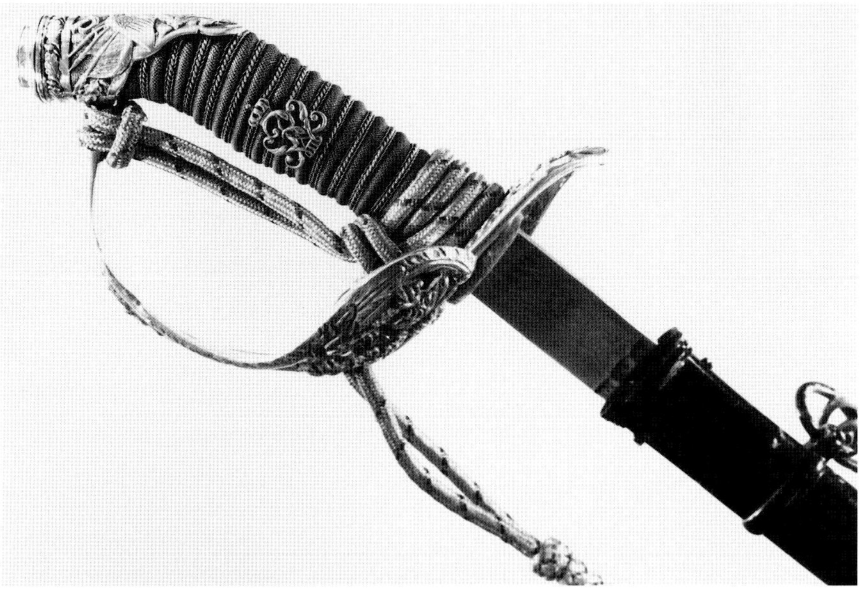

W 153 Infantry Officer Saber M 1889 (for short, I.O.D. 89) in luxurious form. The hilt is richly ornamented with military and floral decorations. The portepee is the one for sea officers. (STJ)

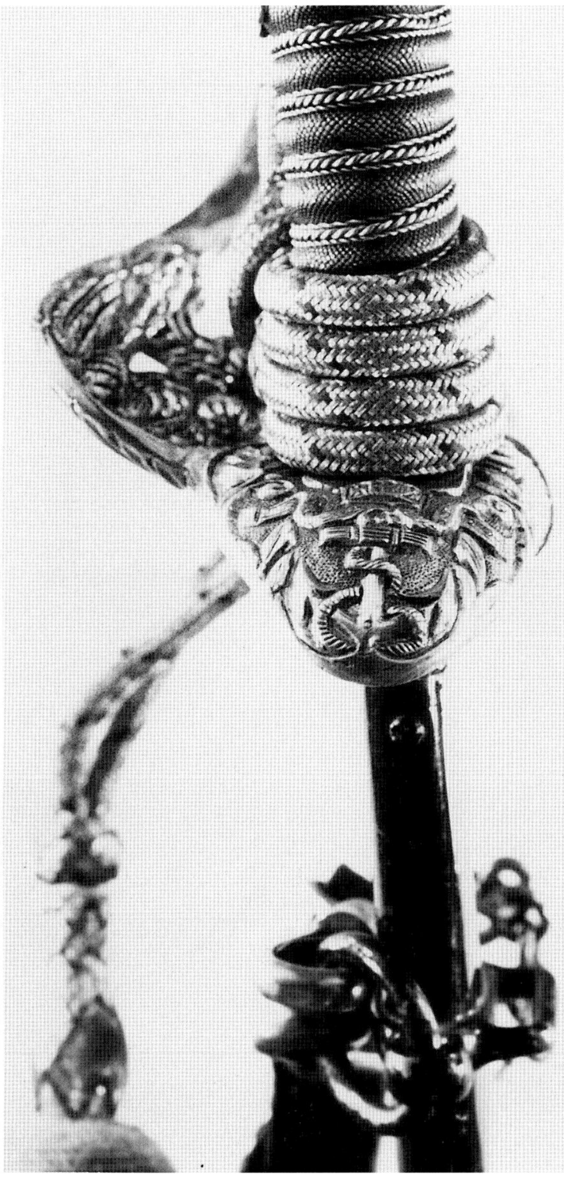

At what point in time the I.O.D 89, also called I.O.D. n/M, was first conceived officially could not be determined. In the *Uniform Regulations for the Officers of the Naval Infantry* of 1898, though, it was allowed with the dress uniform (Section 3.a): *Mounted officers wear high boots for riding and may attach the Infantry Sword n/M to the naval belt.* According to Section 4, it was also allowed for the undress uniform.

Blade: The blade of the weapon has one blade and two hollow channels on each side. The point lies in the middle area between the channels.

Hilt: The hilt consists of—gilded—brass, and is composed of a hand protector, so-called half grip cap with pommel screw, grip, and smooth grip ring. The following elements form the hand protector: grip or main bow; two terz bows; and oval guard. Between the main bow and the upper terz bow is the Prussian eagle facing the "heraldic left" with the royal monogram WR II in the breastpiece, in the crown of which the lower terz bow ends. The handle—cord underwrapping, fish-skin covering, and silver wire overwrapping—is usually made with the crowned royal monogram on the outside.

Scabbard: The steel scabbard has two carrying straps with moveable rings. Until 1905 it was plain, then dark according to an order of June 15.[87] The lower ring (with ring strap) was omitted by order of November 3, 1910.[88]

Like the naval saber, the I.O.D. 89 was also made according to taste and pocketbook—plus the courage to face one's superiors with an extravagant version. Therefore it is not surprising that there was obviously also a special version for officers of the Naval Infantry, which they could buy if they wished.

W 154 The upper side of the guard in detail. This weapon (W 153) has a special feature, an unclear anchor under an Imperial crown (with pendilien). Thus this weapon is undoubtedly assignable to a member of the Naval Infantry. (STJ)

A maritime version of the I.O.D. 89, which will be described, corresponds in its basic form to the prescribed model. Yet the formation of the weapon is elaborate and especially splendid. The handle cap is adorned with military trophies, in front of which is the Prussian eagle with the crown WR II on its breast. The closing on top forms oak-leaf decoration around the grip cap. The pommel screw is wreathed with laurel all around. The bow is finely engraved. On the back of the guard—and easy to see—is an unclear anchor with an Imperial crown set over it and fitted with bandeaux. Under the guard a mussel and two dolphins are portrayed. This rare variant of the I.O.D. 89 for the Naval Infantry was long unknown, even to veteran collectors.

The Artillery Officer's Saber

By order of December 4, 1899, it was required that a naval field battery[89] was to be detached from the Kiauchau Naval Artillery Detachment and attached to the III. Sea Battalion.[90] This battery, by order of June 25, 1900, was to wear the uniform of the III. Sea Battalion with minor changes.[91] Thus the naval infantry again had an artillery component after the year 1867.

Five artillery officers transferred from the Army to this new unit of the Imperial Navy.[92] Surprisingly, the old regulations of the Prussian Navy were taken up again for these officers, namely that they were to wear not the naval saber but the artillery officer's saber—of the land army.[93] Note also photo 32 on this page.

Of course the officers had to equip themselves, and thus they had the opportunity to purchase and wear an artillery saber of their choice. It is not possible to look at all the possible variations, so one weapon will be described as an example.

Blade: The blade usually chosen at that time was a reinforced type with a brace.

32 *A naval field battery officer in charge of firing has left his artillery officer's saber leaning against the wall of the gun position. MOV Archives*

W 155 The artillery officer's saber. This model was also worn by those of the naval field battery who had the right to wear it. (PS)

Hilt: The hilt is made of (gilded) brass. The two half-round parrying bar tabs are decorated on the outside with two crossed cannon barrels. The grip, consisting of a wooden core, is covered with fish skin and wound with two or three silver cords. The grip cap is decorated at the top with a lion's head, and its back as well as the grip-cap tabs are decorated.

Scabbard: The scabbard of sheet steel has two carrying straps with rings of brass, or later, iron. Until 1905 it was plain, then according to an order of June 15 it was dark. The lower ring (with ring strap) was dropped by order of November 3, 1910.

The Horse Doctor and Horse Officer Swords

After the Naval Battery, a Fifth (mounted) Infantry Company of the III. Sea Battalion was established by order of March 29, 1902.[94] Both newly established units were assigned veterinarians in the budget.

Their lowest rank was that of the Unterrossarzt (Junior Horse Doctor), who then still belonged to the group of underofficers with portepee. As Rossarzt (Horse Doctor) and in the higher ranks they were military officials with officers' rank. In 1903 the horse doctors were designated Veterinarians, and as of 1910 they moved from the status of an official to that of an officer. For the horse doctors belonging to the Navy, as long as they were not to wear the tropical uniform of the III. Sea Battalion, essentially the uniforms and sidearms of the animal doctors of the Prussian Army were prescribed.[95]

At the time when the junior horse doctors or junior veterinarians were still portepee underofficers, a sidearm, the horse doctor's saber,[96] was established.[97] This model (ordnance weapon) will be described first.

Blade: The curved blade has a hollow groove on each side. The back is curved concavely before the point (Pandur point).

Hilt: The hilt is made of brass. The two angular parrying-bar tabs are left smooth and have an engraving along the edge, almost forming a square. The grip, consisting of a wooden core, is covered with fish skin and wound with two twisted silver wires. The grip cap is decorated with a lion's head, plus floral decorations in the back and on the grip-cap tabs.

W 156 Horse doctor's saber, (Prussian) ordnance weapon, for lower-rank horse doctors or veterinarians to wear in service. (PS)

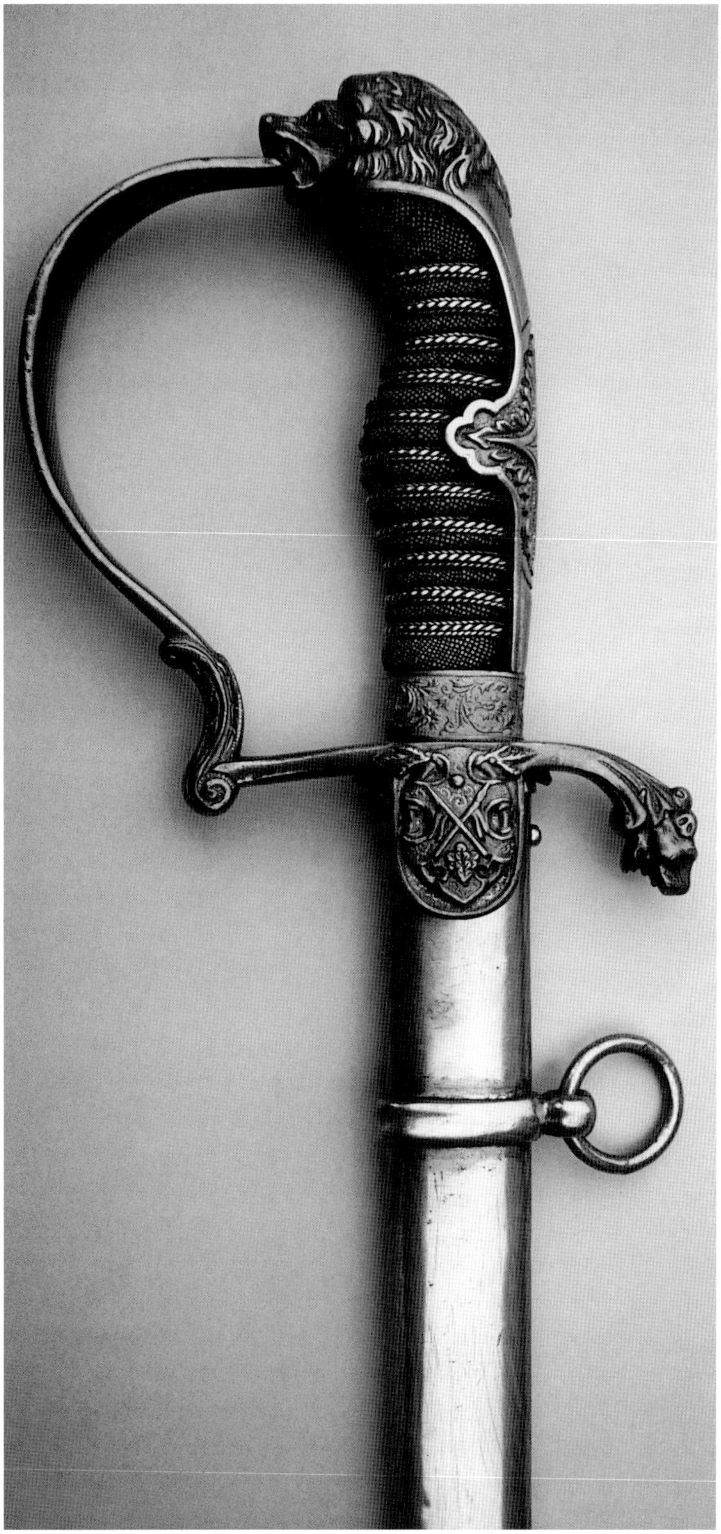

W 157 Cavalry Officer's Interim Saber. This weapon was worn by the animal doctors as military officials or later as officers. There are numerous variations of the model. (PS)

Scabbard: The cast steel scabbard has two carrying straps with brass rings. Until 1905 it was plain, then dark by order of June 15.[98] The lower ring with its strap was dropped by order of November 3, 1910.[99]

As military officials, or later as officers, the animal doctors had to equip and arm themselves. They had to obtain their weapons themselves and thus had the possibility of buying and wearing the Cavalry Officers' Interim Saber.[100] Here too, it is impossible to describe all possible variations, so only one model will be described as an example.

Blade: The blade, as a rule, is a reinforced blade with a brace.

Hilt: The hilt is made of brass. The two half-round parrying bar tabs are decorated on the outside with two crossed sabers, lances, or both. The grip, consisting of a wooden core, is covered with fish skin and wound with two or three twisted silver wires. The grip cap is decorated with a lion's head, plus floral decorations in the back and on the grip-cap tabs.

Scabbard: The sheet-steel scabbard has two carrying straps with brass rings, or later of iron. Until 1905 it was plain, then dark by order of June 15. The lower ring (with strap) was dropped by order of November 3, 1910.

Weapons of Honor

Weapons of honor are weapons, namely swords and sabers, with which officers are honored for special achievements, when a medal or similar gift cannot be given to them for special reasons.[101] Thus, in a German military lexicon of 1869, was the existence of luxurious weapons described, which were awarded by the king, a prince, etc., or also by countries or by an officer corps to officers for special services. A large conversation lexicon that appeared around the turn of the century offers its readers occasions for such recognition:

Weapons of honor, as awards, were and still are often awarded by regimental commanders or rulers. They are, for example, costly swords..., sabers..., boarders' and sappers' axes...etc. The German Kaiser awards swords and sabers of honor for outstanding scientific achievements and shooting...[102]

In another military lexicon the following is added:

...on the 50th or 60th anniversary of deserving generals, it is authorized that the latter receive the like from their subordinate officer corps; princes of the royal house receive sabers of honor from kings on the same occasions.[103]

The following text will deal with this subject.

A Saber of Honor for the Founder of the Prussian Navy

To honor the service and the deeds of Prince Adalbert of Prussia for the development of both the Prussian Navy and the Reich Fleet in particular is not the subject of this book. Undoubtedly he performed great services for the Navy.[104] Although his influence on the fate of the Navy decreased over the course of time for various reasons, his involvement and influence were honored by the Prussian King Wilhelm I on October 29, 1871. On that day he began the celebration of his 50th year of service. The naval officer corps presented to him on this occasion a silver column, to which were attached as decoration the prows of all the warships built during his years of command. From the King he received a saber of honor. In an old newspaper, the title of which is no longer known today, the event and the weapon were described as follows:

His Majesty has honored the celebrant on the high military celebration day with a golden naval saber with genuine golden handle and scabbard fittings. This bears on one side of the Damascus blade the dedicatory inscription: Wilhelm I to Prince Adalbert on the 50th anniversary of his service, on the other the names of the following battles and fights:

Tres Forcas, 7 August 1856 - Sea Battle, 14 April 1864 - Sea Battle near Dornbusch, 24 April 1864 - Nachod, 27 June 1866 - Skalitz, 28 June 1866 - Ko:nigsgra"tz, 3 July 1866 - Colombey, 14 August 1870 - Gravelotte, 18 August 1870 - Siege of Paris 1870-71.

The ornamentation of the scabbard fittings and handle is made in romantic style; on each there is a lion's head with a lavish mane, under this the monogram of the celebrant with a royal crown, while the pierced and decorated basket bears an anchor with an Imperial crown.[105]

Nothing can be reported of the fate of this weapon.

The Saber of Honor of Rear Admiral Deinhard

About *Admiral Deinhard* and his saber of honor there is a little anecdote that shall be repeated on this occasion. Admiral Deinhard had been in service with Reich Commissioner Wissmann in East Africa in 1888-89. On October 31, 1889, he was on his ship, the *Leipzig,* on the voyage home and met the Kaiser's yacht in the harbor of Port Sigri in Mythylene:

> *After the salute and the parade of the ship before the eyes of the Kaiser* (author's note: Wilhelm II) *came the signal: Admiral on board. The admiral had not been prepared for that; hastily he disappeared into his cabin to change his service uniform for his gala one. The clever seaman's courtier did not forget to have him put on the saber of honor given to him by the Sultan of Zanzibar.*[106] *After he reported to the Kaiser, the latter asked him: Well, Deinhard, what do you have at your side? He replied: Your Majesty, when I previously attached my saber, the shoe fell off the scabbard. Then I was compelled to put on the saber given to me by the Sultan of Zanzibar. I beg you most humbly to let me wear it. Laughing, the Kaiser gave him permission. The Kaiser spent several hours on board the "Leipzig" and inspected the ship and the crew. Before he left the ship, he expressed his Imperial thanks to the squadron commander, the officers and men for the sustained and energetic conducting of the blockade and for their achievements in the retaking of the coast.*[107]

The saber of honor that Admiral Deinhard had been given by the Sultan of Zanzibar was naturally not a German naval saber. It was instead a weapon such as was customary in East Africa. A drawing by the naval artist C. W. Allers shows Admiral Deinhard with this saber.[108] According to the picture, this can be identified as a so-called Mameluke saber. Since nothing further is known of the provenance of this saber, no further details of it can be given.

35 Vice-Admiral Deinhard with the Sultan of Zanzibar's saber of honor. C.S. Archives

The Shooting Prize Saber

In the Imperial Navy, weapons of honor were also awarded. They came late, for in the Army weapons of honor were already given for marksmanship in 1888.[109] In 1903 the Navy was also given the possibility of awarding such weapons of honor.

By order of January 24, 1903 Kaiser Wilhelm II declared:

For the best shooting performances of officers and underofficers of my Navy with the rifle, prizes of honor will be given every year in accordance with suggestions made by me. At the same time, I authorize the directives made by me in completion of the Navy's rifle-shooting prescriptions for the competition for the prize of honor and empower them to arrange it further.[110]

For the officers a saber was suggested as the prize of honor, and for the underofficers a pocket watch. The detailed regulations for competing were published in the rifle-shooting regulations for the Navy.[111] There it was stated:

> *214. For the naval units on land, including naval infantry, and at sea, that have conducted school shooting, a competition takes place every year for the prizes of honor donated for officers and underofficers for the best shooting achievements with the rifle. The competition will be carried out in the prize-of-honor shooting.*
>
> *215. Required to participate in this shooting are the first lieutenants and lieutenants of the sea officer corps and the naval infantry belonging to the aforementioned naval units, deck officers of the seamanly personnel and the naval artillery, who are summoned to shooting service, underofficers of the seamanly personnel (exclusively sea ensigns), the seamen's artillery and naval infantry, as well as vice-sergeants and exercise underofficers who have taken part in school shooting, and who are not prevented by unavoidable service, sickness and the like on the days set for the prize-of-honor shooting.*
>
> *Permitted to take part in the shooting are the staff officers, the captain-lieutenants or captains as well as the captain-lieutenants at the command posts, and the not above permitted deck officers of the seamanly personnel and the seamen's artillery of the naval units that come into consideration for the prize shooting.... Officers and underofficers who have already won a prize of honor (in the Navy or the Army) for best shooting performances with the rifle are closed out from further competition.*

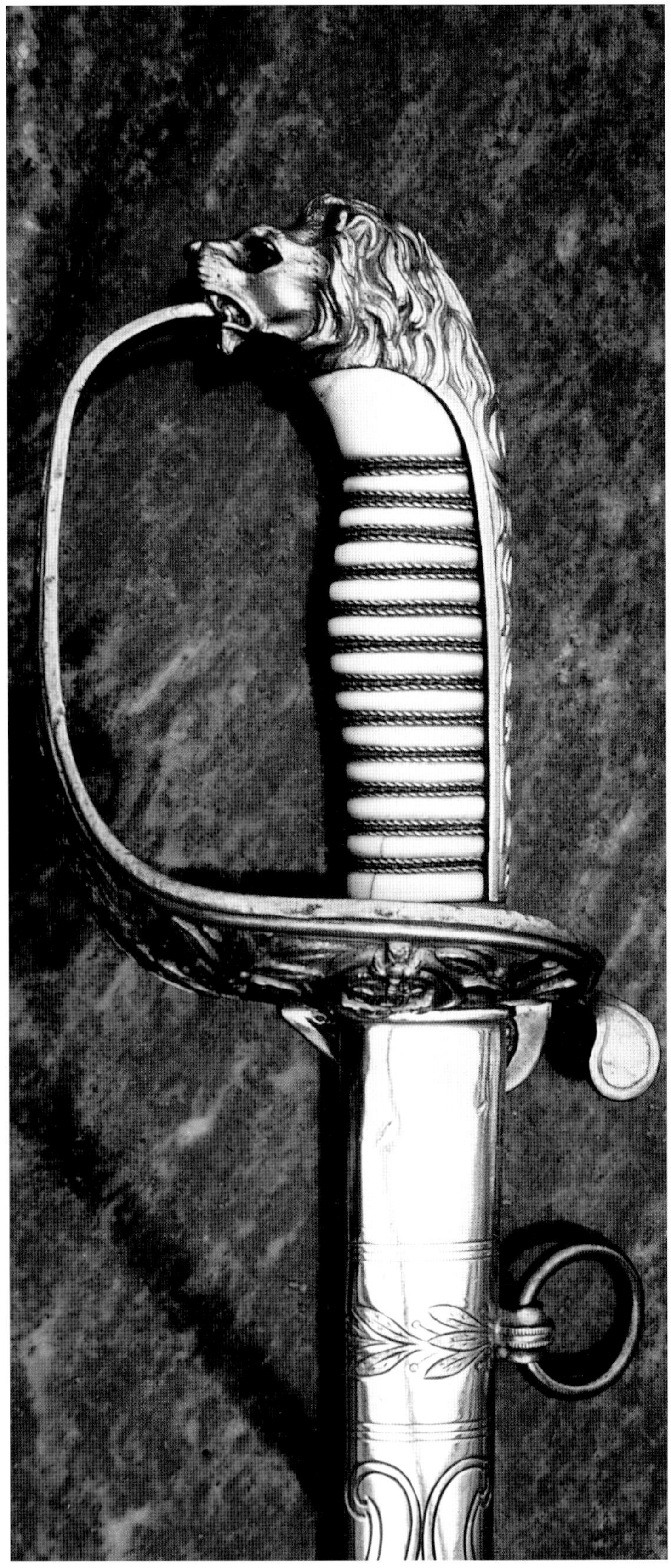

W 158 Hilt of a shooting-prize saber for members of the Navy. Note the laurel decor at the level of the ring holder. (PS)

216. The prizes of honor are presented in the name of His Majesty. They consist for officers of a saber inscribed with the name of the marksman and the reason for the presentation, for deck officers and underofficers of a pocket watch with the same inscription.

Every naval station receives a prize every year for the best marksman among the officers and two prizes for the best two marksmen among the underofficers.

For the naval units in Kiauchau (excluding the cadre formation), the prizes of honor for the Baltic Sea station come into consideration.

217....

218. For the prize shooting the following conditions are established:

Target: Ring target with 24 rings, of which Ring 24 has a diameter of 2.5 cm, while the diameter of the other rings increases by 2.5 cm each time.

Weapon: Rifle of the involved naval unit, but it is always allowed for the officers to use the Rifle 98.... Distance: for Rifle 98: 150 meters, for Rifle 71/84: 100 meters. Number of shots: 7, with 3 lying inclined, 4 lying with free hands.... The prizes of honor are to be distributed only when at least 140 rings are fired.

219. For the determination of the best marksman, first of all the sum of the fired rings is decisive. If these are equal, then the last shot, or if necessary the second last, third last, etc., decides.

220. The fleet command, the commands of independent squadrons or independent divisions, the inspection of the training system, the ship artillery, the coast artillery and the mine system, the torpedo system and the naval infantry, the naval inspections as well as the ship commands directly subordinate to the naval stations will send the involved station commands as soon [p. 106] as possible at the close of the shooting year the names of the best marksman of the officers and the nest two marksmen of the underofficers, with their shooting results.

The station commands will thereupon determine the marksmen who are eligible for the awarding of the prizes of honor and send the names and shooting results to the state secretary of the Reich Naval Department by December 15 of each year.

(Statement of the saber scabbard length required.)

221. On the basis of this information, the state secretary of the Reich Naval Department will send the prizes of honor immediately to the appropriate naval units, which are to conduct their presentation to the best marksmen in an appropriately solemn manner.

W 159 The basket of one of the rare shooting-prize sabers. This early weapon (of 1905) is still decorated with lines on the basket. (WGM Inv.No. 000568)

In addition, the names of the officers and underofficers awarded prizes of honor are to be reported to His Majesty along with the shooting results and be made known in the naval information publication....

The bearers of such naval prize-of-honor weapons are being published for the first time. In all, 22 officers could be identified. Whether shooting prizes were still awarded in the year of 1914 is unlikely, since the outbreak of the war ended activities such as shooting competitions.

List of the Imperial Navy officers to whom weapons of honor were awarded

1903	Kapitänleutnant	Bechtel	North Sea Station
	Oberleutnant zur See	Grumme	Baltic Sea Station
1904	Oberleutnant	Vitzthum von Eckstaedt	Baltic Sea Station
	Leutnant zur See	Plange	North Sea Station
1905	Kapitänleutnant	von dem Knesebeck	SMS Hohenzollern
	Oberleutnant	Hüttmann[112]	III. Stamm Sea Battalion
1906	Leutnant zur See	von Bredow	SMS Prinz Adalbert
	Oberleutnant	Hannemann	II. Sea Battalion
1907	Leutnant zur See	Gayer	SMS Lothringen
	Oberleutnant zur See	Hellwig	II. Seamen's Division
1908	Leutnant zur See	Grothe	I. Torpedo Division
	Kalitänleutnant	Kurtz	SMS Planet
1909	Kapitänleutnant	von Alvensleben	I. Torpedo Division
	Oberleutnant zur See	Baron Röder v. Diersburg II.	Torpedo Division
1910	Kaspitänleutnant	Gautier	I. Seamen's Division
	Oberleutnant	Hermann	III. Stamm Sea Battalion
1911	Hauptmann	von Pape	III. Sea Battalion
	Oberleutnant zur See	Riesel	SMS Gneisenau
1912	Oberleutnant	Klinger	III. Sea Battalion
	Oberleutnant	von Borcke	II. Sea Battalion
1913	Oberleutnant zur See	Frege	II. Seamen's Artillery Unit
	Leutnant zur See	Raspel	SMS Hannover

W 161 Terzside inscription of the blade of the W 159 saber (page 105). (WGM Inv.No. 000568).

The prize-of-honor saber for the Navy followed the ordnance-type pattern of the naval officer's saber, but varied from the standard model in its decorations and gilding. The diferences will be described below. The prize-of-honor sabers for the Navy, unlike those of the Army, were not marked with acceptance stamps.[113]

Blade: The blade is made of high-value Damascus steel with the Eisenhauer mark of quality. It is about two-thirds gilded, with a panel on the outer side of the blade etched with black-shadowed gold lettering in two lines:

Prize of Honor for outstanding shooting achievement... [followed by the year]

to... [bearer's rank and name].

Floral decorations and crossed flags border the text at the top and bottom.

On the inside of the blade are three groups of symbols, surrounded by floral etchings. In the cartouche nearest the point of the saber, in high relief on a blue background, a gilded unclear anchor is etched; in the same manner at the middle is a W II monogram with an Imperial crown right above it; and nearest to the hilt is a ship, and somewhat deeper is the Damascus steel designation. On the back of the blade, at the level of the dull edge, the manufacturer's name: Weyersberg, Kirschbaum & Cie, Solingen, is etched.

Hilt: Between 1903 and 1907 the form of the hilt was changed several times.

An early hilt (of 1905) has the form of the ordnance weapon, though the appearance is fancier and the workmanship is finer. The fields between the line decor on the front part of the bow are decorated with raised oak leaves. The back part, on the vother hand, was decorated with laurel. The bottom of the flap was also decorated with laurel. When one views the medallion with the Imperial crown and anchor, one sees that here too the empty space is ornamented with oak leaves to the rear and laurel to the front. The hilt was finely gilded.

On a later hilt, the gilded basket is all decorated with raised laurel leaves. The line decor was omitted, as were the decorative elements in the medallion.

The medallion with the Imperial crown, and the unclear anchor on the basket was like that of the ordnance weapons.

Scabbard: The scabbard material is leather, but sheet iron covered with leather was also used. The fittings correspond to both versions of the ordnance weapon, but are also decorated with engraved laurel at the level of the carrying rings and on the mouth fitting.

The Graduate Saber of the Friedrich Wilhelm Institute

In order to assure the medical care of the Army and Navy to a sufficient degree, the medical corps was enlarged in three ways:

1. With graduates of the military training institutions,
2. With doctors who were serving their general service duty (one-year volunteers),
3. With such doctors as entered the Army for long periods.

The military doctors' training institution, opened as the *Pepiniere* in 1795, was named the *Friedrich-Wilhelm-Institut* in 1815 and was known since December 2, 1895, as the *Kaiser-Wilhelms-Akademie* for the *Military Doctors' Training*. This institute offered thorough training in all areas of medical science in common with the medical school of the University of Berlin.[114] Along with medical training, the trainees also received basic military training. After passing the state examination, they were inducted into the Army or Navy at the rank of Unterarzt. The Navy took part in the institute's work since 1863 by assigning an Oberarzt, plus a second as of 1869. Since 1869 so-called Naval Cadets—eight per year—were also accepted for training.[115]

By the end of the 18th century it was already customary for the director of the institute to present to every graduate, in a public ceremony, a sword (for Army members) as a memento of the institute. The sword bore on its blade the name of the institute. Under special conditions, students could also obtain the weapon as an honor after three years of study.

W 164 Hilt of a Friedrich-Wilhelm-Institute graduate's saber for naval doctors. The form of the weapon corresponds to early Imperial naval officers' sabers (compare W 64, page 115). (PS)

After naval men had completed the training successfully, the naval officer's saber was presented to them. The graduates' sabers were not particularly decorated, but largely corresponded to the ordnance pattern, differing from that and from each other in details. On the quartside flap the name of the institute, *Königlich Medizinisches Chirurgisches Friedrich Wilhelms Institut,* was engraved in abbreviated writing. In a narrow sense the graduates' weapons do not belong among the weapons of honor. But they are mentioned at this point because of the "honorary character" that they have.

33 A Naval Senior Staff Doctor with the naval officer's saber. D.S. Archives

W 165 Inscription on the flap of the graduates' saber (W 164). (PS)

W 166 Inscription on the flap of another graduate's saber. The hilts of the weapons are identical, but the inscription varies slightly. (MDG)

The Saber of Honor for the II. Sea Battalion

In the II. Sea Batatllion a pleasant custom had taken root, namely that members retiring from the officer corps, if they wished, were presented with a finely worked saber. On the basis of the weapons still to be seen today, they were based on a non-allowed model, a saber with a pierced basket. One may note the weapons of Oberleutnant Staab (W 167) and Leutnant Herdemann (W 95).

Section 122 of the uniform instructions for officers of the Naval Infantry[116] states as to bearing appropriate weapons:

Swords of honor, that were were presented or granted by an officer corps, plus swords that were given by members of the Imperial house or other royal houses, may be worn—in a leather scabbard and on the prescribed belt—without special permission.

The officers who returned to the Army often had the wish to go on wearing these weapons at least out of service. But it can be presumed that they had great difficulties with their commanders if they wore the naval saber there. This may be assumed, although the regulation cited above applied almost word-for-word to the Prussian Army, with the exception of the leather scabbard prescribed for the Naval Infantry, as the Army required a steel scabbard.[117] It was more difficult for officers from Bavaria, Saxony, and Württemberg, since their kings could make the regulations for their uniforms. Thisproblem was solved only in 1913.

W 167 Hilt of a naval saber presented to a member of the II. Sea Battalion by the officer corps of the Naval Infantry on the occasion of his return to the Army. On the handle is the star of the Order of the Black Eagle.

W 168 The pierced basket of the naval saber. Obviously, this non-allowed version was freely presented to departing officers of the II. Sea Battalion (see also the pendant of W 95, page 75). (PS)

In the II. Sea Battalion, as also in the Ist, several high-ranking personalities of the time were placed "a la suite."[118] One of them was *Colonel Egon Prince of Fürstenberg*,[119] Oberstmarschall of the German Kaiser and a personal friend of Kaiser Wilhelm II. In 1912 he presented to every officer of the II. Sea Battalion—with the Kaiser's permission—a saber of honor. From then on every officer who departed from the battalion was also to receive such a weapon.

To obtain official permission to wear the saber of honor, the Comander of the II. Sea Battalion wrote to his superior office as follows:

W 169 Blade inscription of W 167: Dedication from the officer corps. (PS)

W 170 Blade inscription of W 167, stating the time of service, place and ship. Leutnant Staab was a training officer on SMS Victoria Louise. (PS)

Wilhelmshaven, 3 July 1913

Command of the II. Sea Battalion B. No. 2182 I.

Re: Saber of Honor and Flag of Arms

On the occasion of the presence of His Serene Highness the Prince of Fürstenberg with the Battalion in the previous year, His Serene Highness presented a saber of honor to every officer who then belonged to the Battalion and who would leave the Battalion in the future.

34 Max Egon, Prince of Fu:rstenberg, a la suite of the II. Sea Battalion, who presented sabers of honor to the officers of the II. Sea Battalion. Fürstenberg Princely Archives

W 171 Hilt of the memorial saber. On the handle is the coat of arms of the Fürstenberg family. Instead of an Imperial era hilt, the pattern was that of the 1923 model (see also p. 123 and p. 140). (FFM)

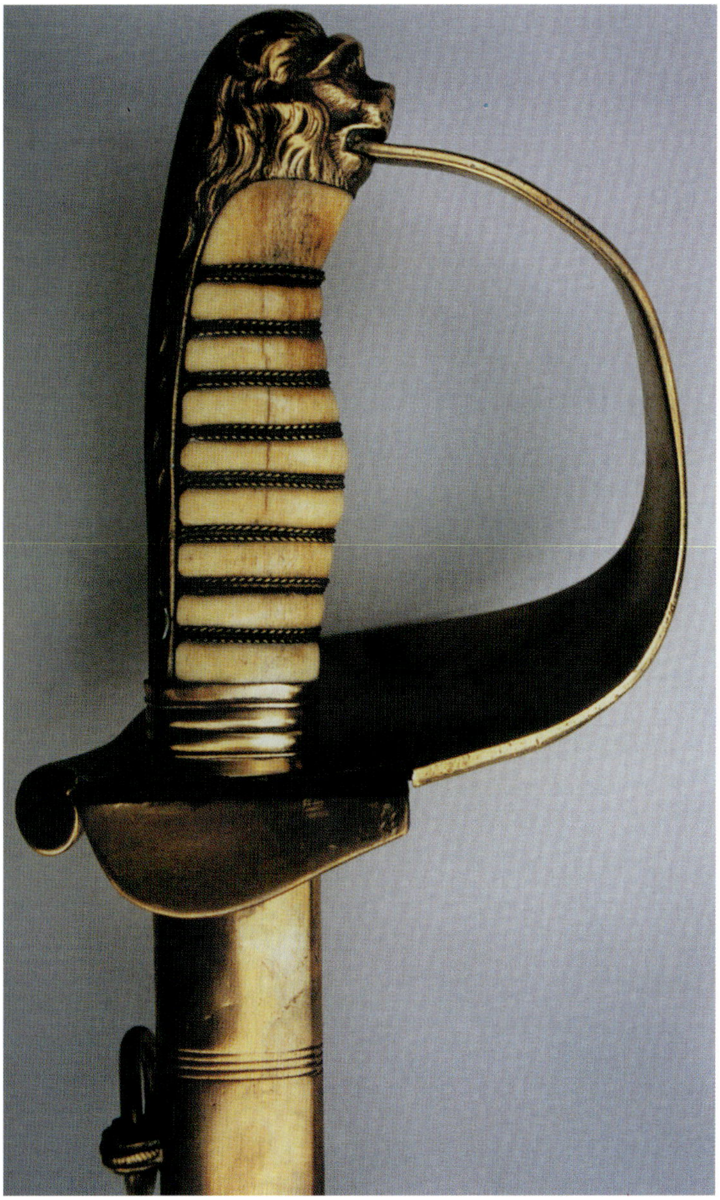

W 4 Hilt of a Prussian naval saber with the small lion's head: the flap, as usual with this model, has no hole to hold the pin soldered onto the scabbard.

W 11 Hilt of a naval saber (see also W 8, page 21) with a black handle: this version was planned for deck officers and portepee underofficers in the first years of the Prussian Navy. The material of the grip is horn, but—we have been told—grips with dark ray skin are also said to have existed. (PS)

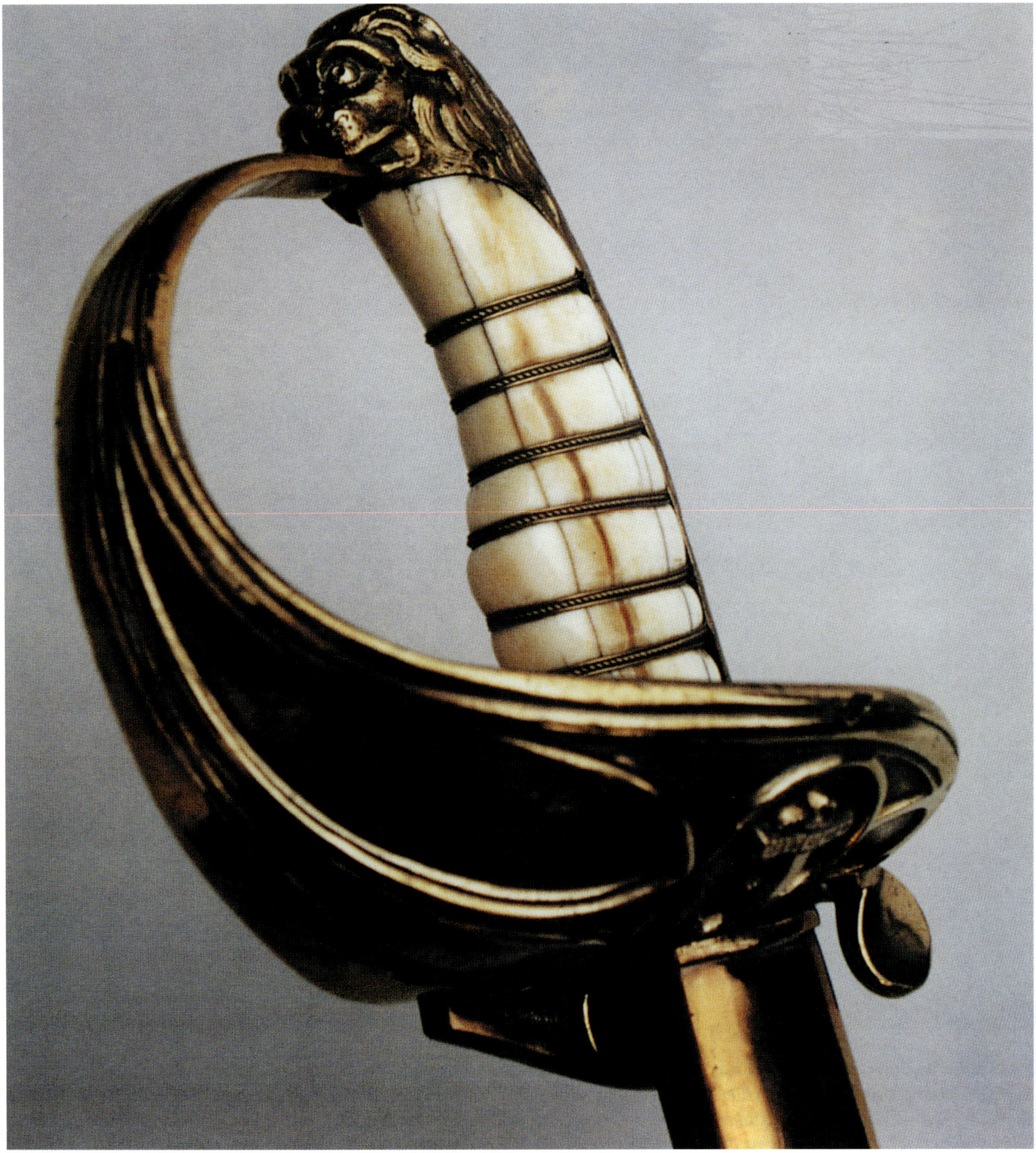

W 6 Hilt of a Prussian naval saber. The lion's head has just the indication of a lower jaw. (PS)

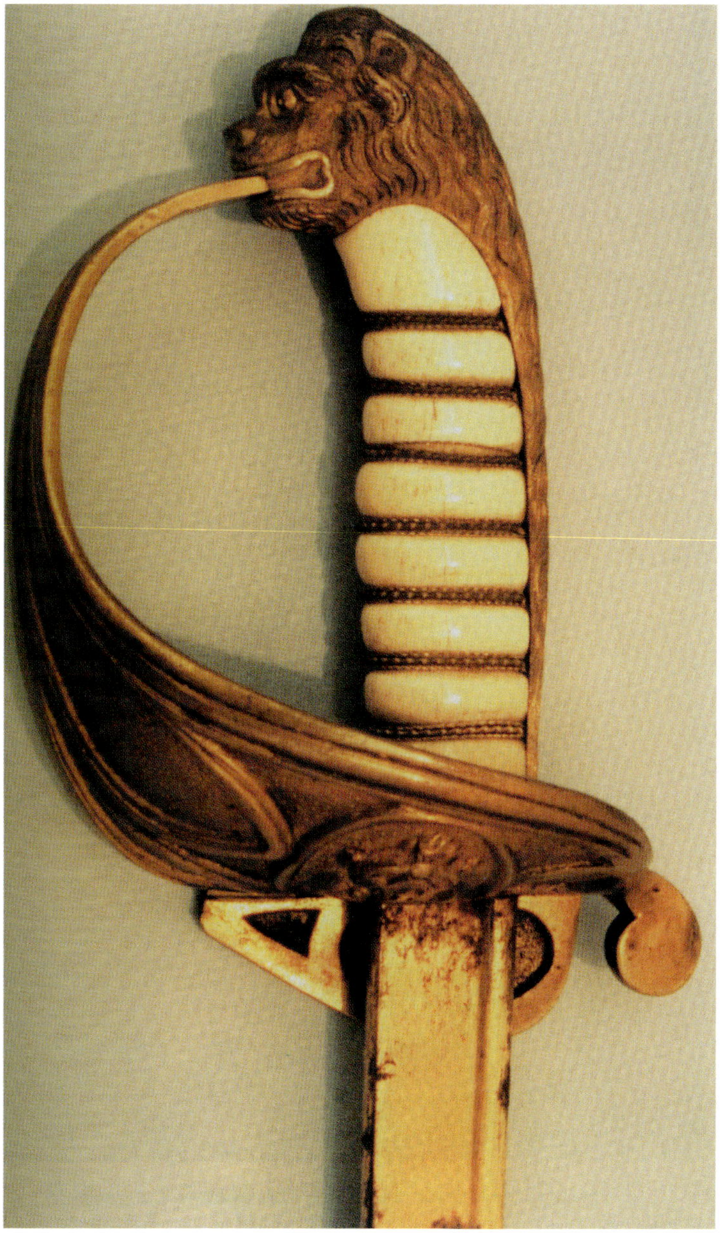

W 40 A very early naval saber for a member of the Imperial Navy (circa 1872). All visible features correspond to those of the Prussian naval saber of 1849. (SPT)

W 64 Hilt of an early naval saber of the 1876 model (privately owned). This weapon still has few grip windings, like the naval saber of 1849. (PS)

W 36 Close-up of the basket of another Schleswig-Holstein naval saber. The large unclear anchor and the edge decorations that end as a sea serpent correspond to the decorations of a deck officer's saber of the Reich Fleet. (PS)

W 29 Deck officer's saber from the 1849 Reich Fleet, grip cap without tabs, grip with ray skin. (PS)

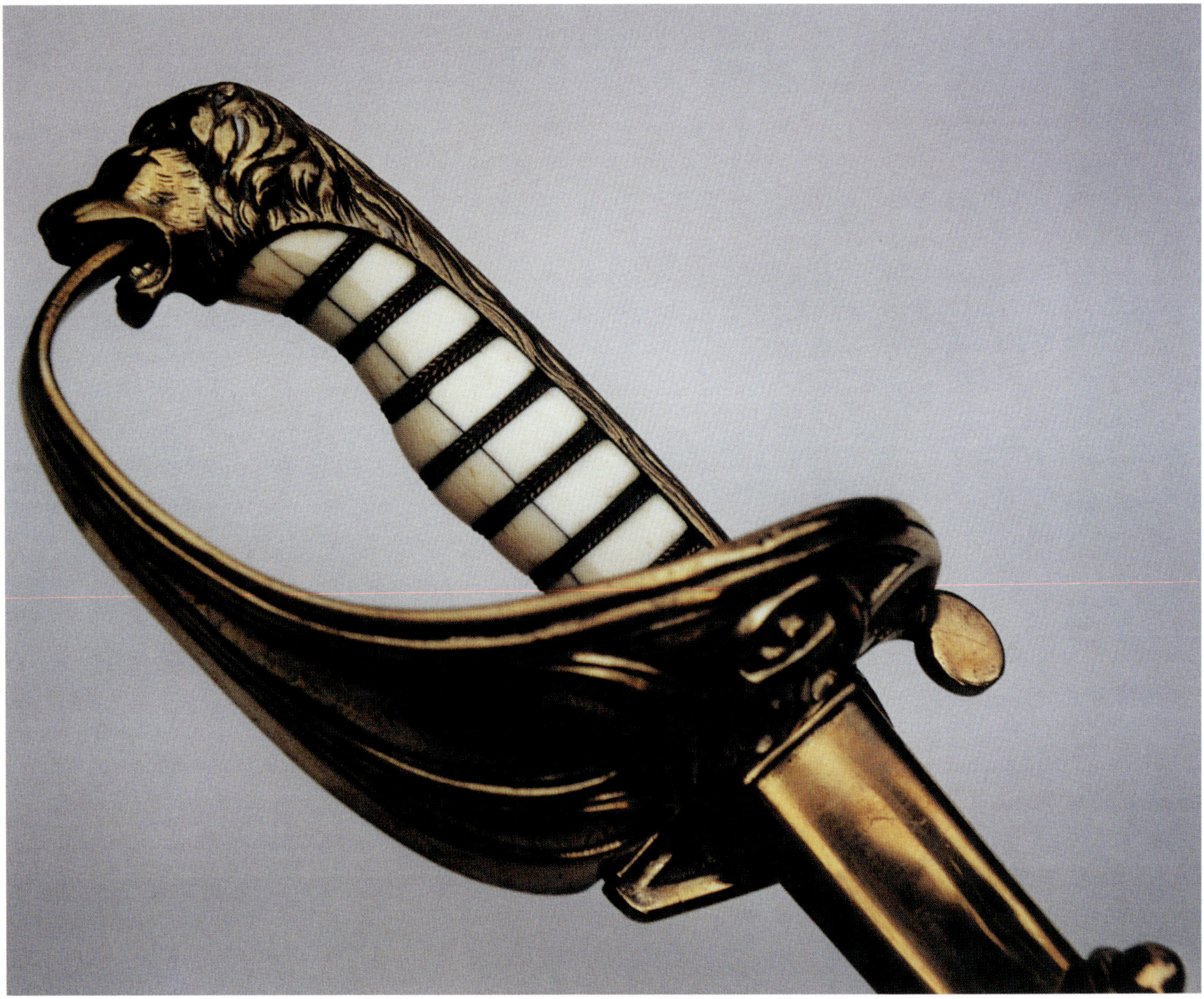

W 43 Another naval saber of 1874. (PS)

W 103 Imperial naval saber, special version. Instead of the otherwise customary lion's head, this weapon has an eagle's head. This was not authorized at the time, but was tolerated. Eagles' heads on naval sabers are very rare. The basket has a trapezoidal flap. (PS)

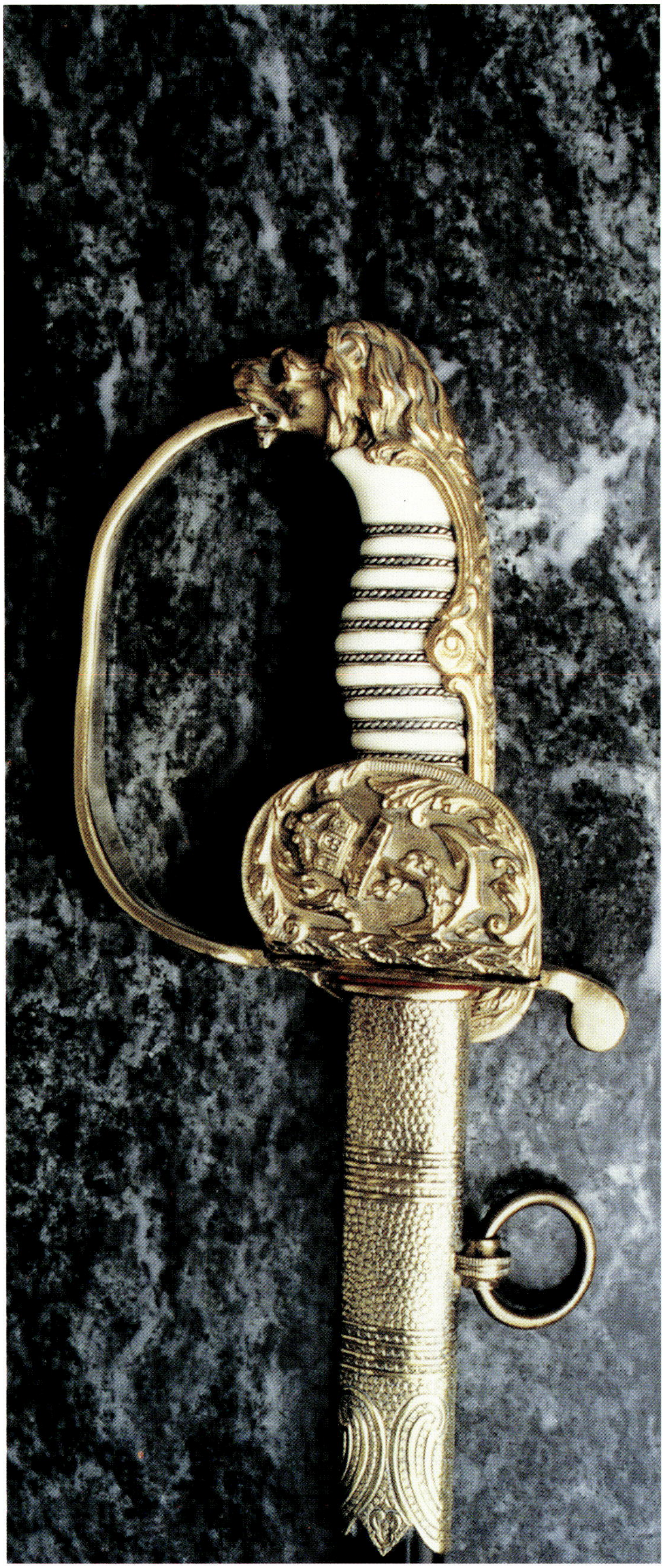

W 113 Imperial naval saber, similar to W 111 (page 81). Instead of the laurel decor, this weapon has arabesques. Note the especially finely worked scabbard. (PS)

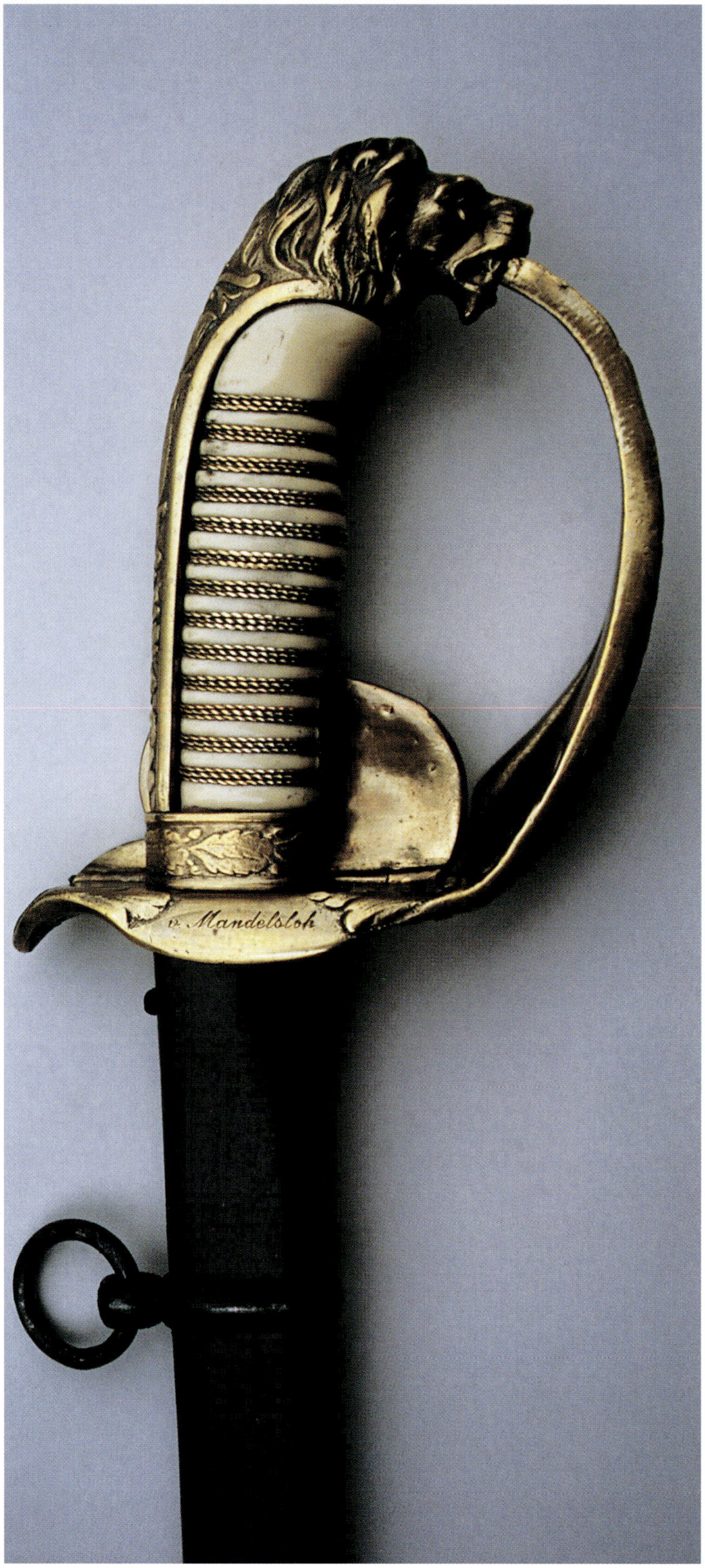

W 121 Back of an Imperial naval saber. This weapon has no flap, instead the quartside guard curves downward. Through the leaf decor that ends at the grip level, the guard has a plain area which was often used for engraving the owner's name, which is von Mandelsloh here. The weapon has a steel scabbard, as sometimes seen with sabers of Naval Infantry members. (PS)

W 192 Basket of a naval saber, luxuriously made, with raised oak-leaf decoration. (PS)

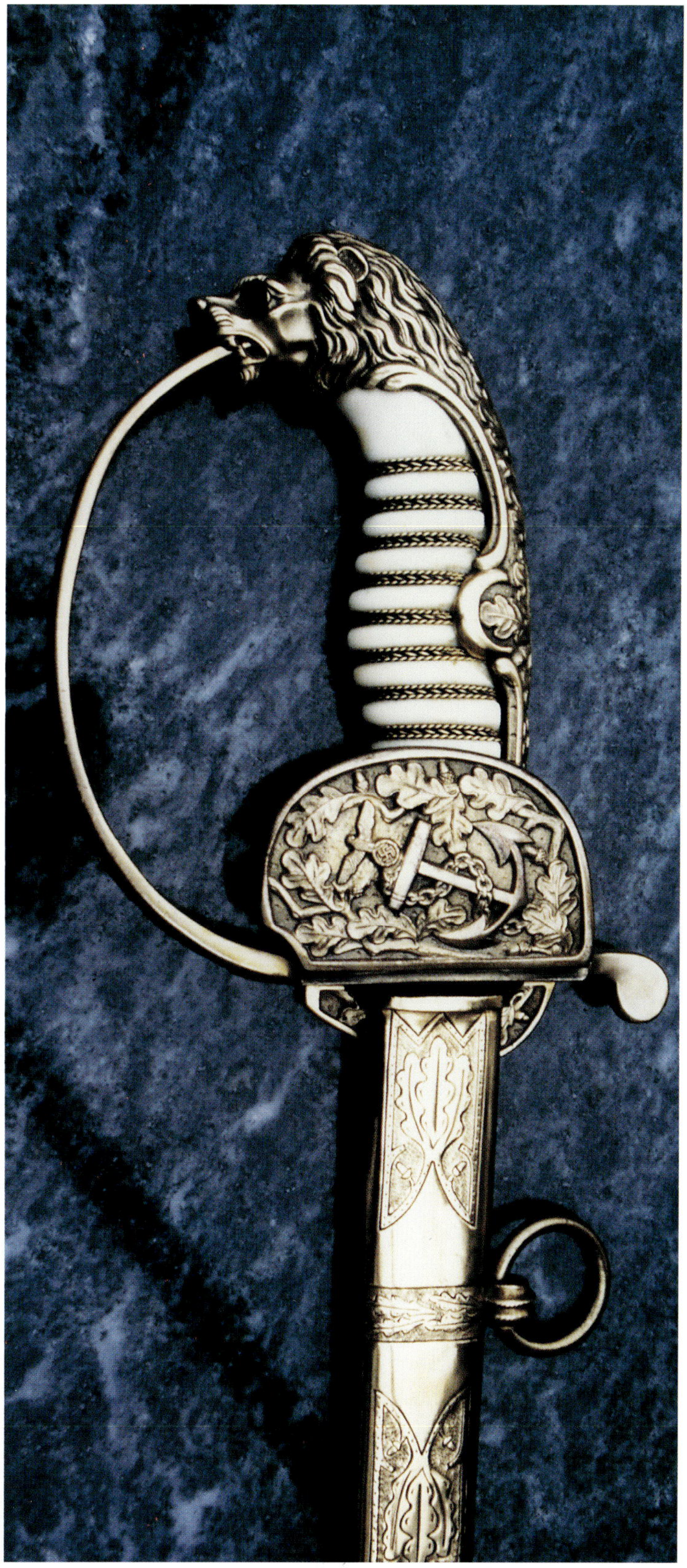

W 195 Hilt of a naval saber of Kriegsmarine times. This very rare weapon corresponds to the pattern in the catalog of Anton Wingen. Note also the hand-engraved scabbard with oak-leaf decoration in the style of the times. These scabbards were made only to order on account of their cost. (PS)

W 160 Basket of a rare shooting-prize saber. This weapon (of 1907) is decorated only with laurel leaves on trhe basket. (PS)

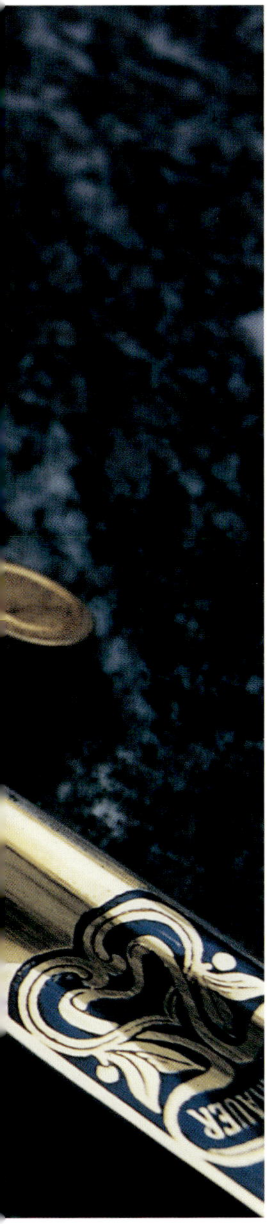

W 181 Naval (officer's) saber of the Reichswehr, from 1923. This (ordnance) weapon was supplied to officers in service. The portepee is made of spun aluminum. (PS)

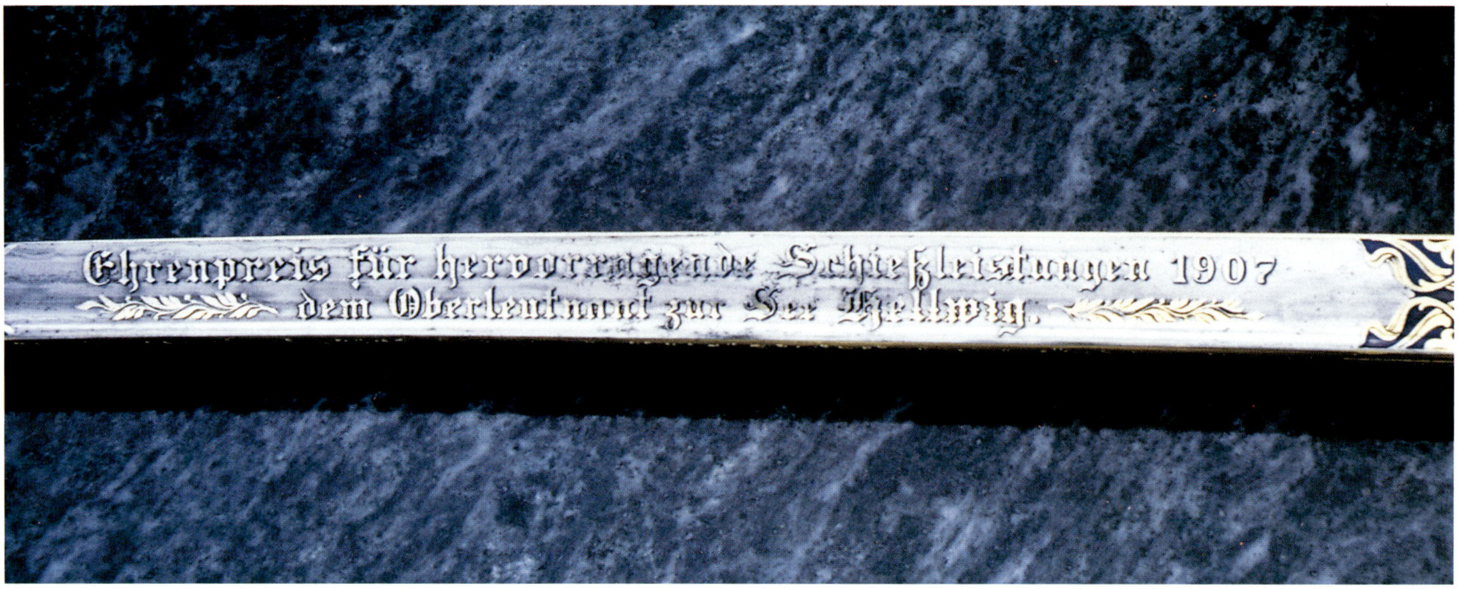

W 162 Terzside blade inscription of the W 160 saber on page 122. (PS)

W 163 Quartside blade inscription, (slmost) identical on both shooting-prize sabers. (PS)

W 193 Blade of a saber from Kriegsmarine times (close-up). Note the portrayal of the ships. The flags on the masts were decorated with small swastikas. (PS)

W 190 Basket of a luxuriously made naval saber. Around the medallion, the basket is decorated with laurel leaves. All the scabbard fittings are also peened and have decorative bands. These bands have a meander pattern composed of small swastika arms. This type of scabbard decoration is very rare. (PS)

W 196 Basket of the W 195 naval saber from the Kriegsmarine era. The royal eagle, facing eastward, was worn on naval uniforms as of March 24, 1934. (PS)

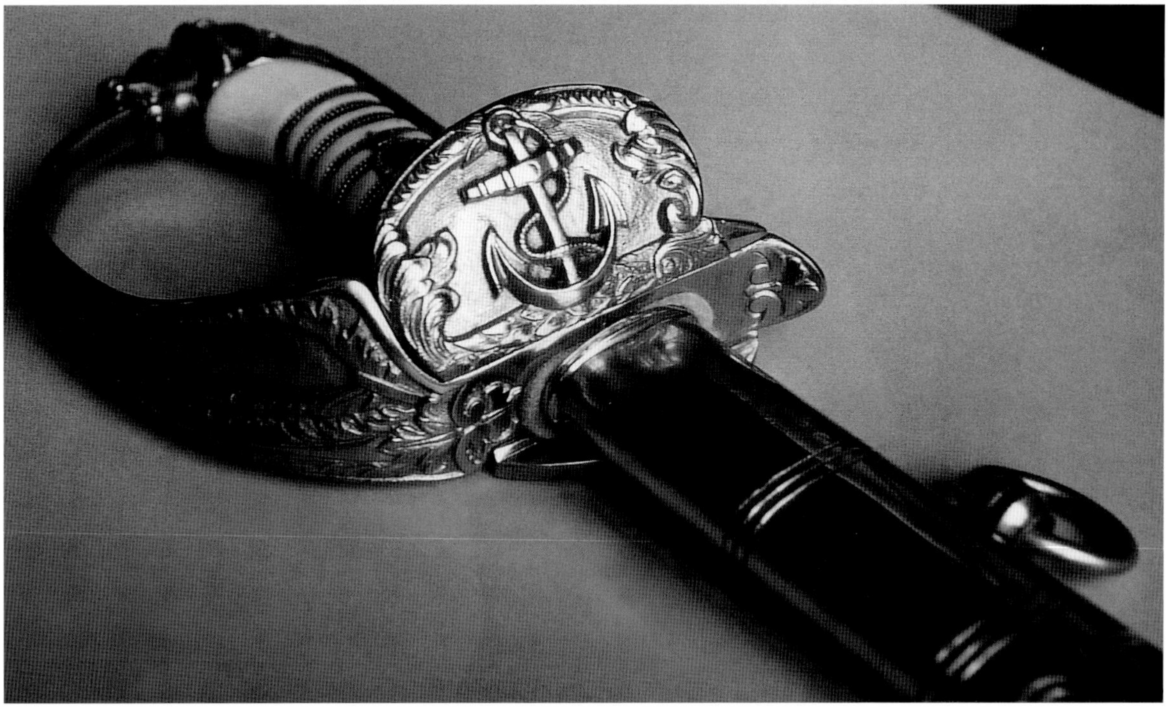

W 172 Basket of the W 171 saber, page 112. (FFM)

His Serene Highness likewise presented two flags of arms for the trumpets of the battalion:

On this occasion His Serene Highness said that His Majesty had given oral permission to award the saber of honor and the flags of arms. An official confirmation of the All-highest permission has not reached the battalion.

Since it is the great wish of the officers of the Battalion also to be permitted to wear this sword of honor later in the Army, the Battalion requests the permission of Station N. to add to the acceptance of the gifts and the All-highest permission thereto and to authorize the wearing of the presented weapons. The sabers, aside from inscriptions and decorations on the blades and hilts, correspond to the pattern prescribed for officers of the Navy.

A drawing of the flags of arms is enclosed.

To the Imperial Inspection of the Naval Infantry Kiel A 6517/B[120]

(signed) von Lettow

The application was finally presented to the State Secretary of the Reich Naval Department. Admiral Tirpitz took up the matter on October 8, 1913, and asked the war ministries of Prussia,. Bavaria, Saxony, and Württemberg to obtain their permission of their sovereigns or ministers of war. After disposing of several concerns, the permission of Prussia[121] was gained on November 7, that of Saxony[122] on November 26, and of Bavaria[123, 124] on December 3, 1913.

Such a saber, presented by the Prince of Fürstenberg, could not be located to date. But there is a saber in the Fürstenberg Princely Museum (W 171) which bears the dedication on the blade:

Maximilian Egon Prince of Fürstenberg to the Officers of the II. Sea Battalion.

W 173 Terzside blade inscription. (FFM)

W 174 Quarterside blade inscription. (FFM)

43 Admiral Valois, wearing the hooked-on naval officer's saber. MSM Archives

44 A naval staff pharmacist with a naval saber. D.S. Archives

45 A lower military official (machinist of shipyard vehicles) with a naval officer's saber. Huss Archives.

This specimen, which has the hilt of the 1923 naval saber, may be a later production version. After World War I the officers of the Sea Battalions maintained the tradition in the Naval Infantry Officers' Organization of Hamburg until the 1970s. After World War II, as far as is known, at least two memorial sabers were made with the permission of the Officers' Organization. One of these weapons is portrayed here.

The Naval Saber in World War I

36 A deck officer in World War I with iron-mounted naval saber.

At the beginning of World War I, the naval officers and underofficers with portepee wore "their" sabers, as far as the prescriptions for the various types of uniform allowed it.

Soon, though, it turned out that the need for (handguns and) sidearms was so great that the weapons industry could no longer keep up with orders. The War Ministry reacted at once. Firms that had formerly produced exclusively civilian goods were encouraged to produce war materials. For the Army it was ordered, among other things, that captured weapons be gathered and turned over for use in training schools and home bases.[125] Since the precious metals, and especially the colorful metals that were needed for the production of particularly important war materials, such as the control rings of grenades, had become scarce, the War Ministry finally ordered that these metals be replaced by types of lower quality:

> *With increased orders for the Army and Navy, the demand for such metals, which are not found, or found in only small amounts, in the homeland, has increased steadily. In view of this fact it is regarded as one of the most important tasks to convert, immediately and as far as possible, to substitute materials, and to transport old material of the specified type to appropriate places as quickly as possible.*
>
> *The chief metals (including graphite) that come into question are listed here:*
>
> *I. The group of metals whose use is to be limited as much as possible (conserved group.)*
>
> *1. Nickel*
>
> *2. Tin*
>
> *3. . . .*
>
> *6. Copper*
>
> *II. The group of metals that are to be used when possible as substitutes (substitute group).*
>
> *1. . . .*
>
> *3. Steel and iron*
>
> *4. . . .*

Within the conserved group I, the metals are listed in order of their immediate significance, so that even the substitution of metals with higher numbers (such as copper) is to be regarded as a gain and striven for, naturally provided that a complete replacement by a metal from Group II is impossible.

There follow some examples for clarification:

Replacement of brass, copper and nickel . . . by zinc-plated iron . . .

For the purposae of attaining the indicated goal, it is determined:

1. All delivery prescriptions are to be tested and changed according to the reaplceability of the metal parts in the above sense. . . .[126]

On the basis of this—formerly secret—order,[127] *also called Metal Saving Order I, the production of bladed weapons for the Army and Navy was gradually halted. By order of October 4, 1915, the War Ministry finally issued unmistakable instructions:*

During the war, conserved metals may not be used for uniforms or equipment. In their place, zinc-plated iron is to be used....[128]

These orders also applied to goods for the Navy. Now the still-needed sabers were made of iron, and at first they were given a thin plating of brass, which was replaced later with black lacquer.

During the course of the war the long sidearm lost significance for the Army and Navy, not least because of the positional trench warfare.

Finally it was generally ordered, including for the Navy:

His Majesty the Kaiser and King has deigned to order that in the theater of war, by all officers of the combat troops, including regimental commanders—but not by the officers of the higher staffs—a short, erectible sidearm instead of the long officer's sword or saber be worn, and in fact

a) by the airship and aircraft officers the [p. 131] short sidearm 98 (with saw back) introduced for the men of these troops.

b) by all other involved officers, the sidearm 98/05 (with saw back) introduced for the men of the foot artillery and engineers.

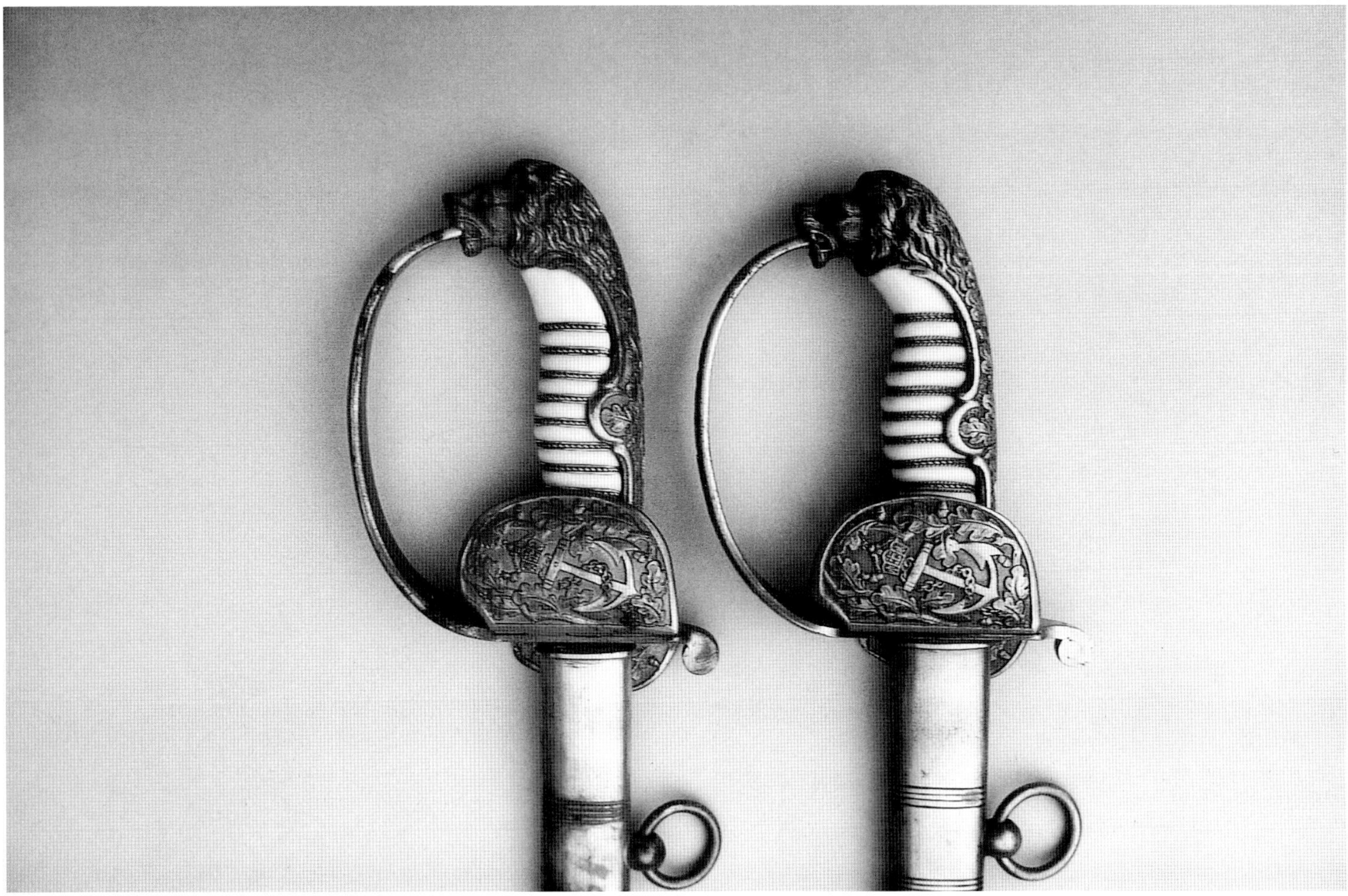

W 175 Hilts (terzside) of two naval sabers from the time after 1916. Both weapons are made of iron. The one at left stil has a brass coating. (PS)

The sidearms are to be carried in the men's sidearm pocket on the leather belt worn in place of the field belt. The attachment of the portepee is to be done according to the regulations for the saber swordknot of the men....

The directive likewise applies to the underofficers who wear the officer's sidearm. The underofficers are to carry the short sidearm in the men's sidearm pocket on a body belt over the uniform jacket or coat.[129]

This order meant an extensive retirement of the naval saber for the duration of the war.

In 1917, by a further order, the limitations on wearing long bladed weapons were finally applied to the higher staffs and also applied to the Navy, for example, including their units in occupied territories:

His Majesty the Kaiser and King has deigned to order that, also by the officers of the higher staffs from brigade commanders inclusive upward, in the combat area, the short sidearm will be worn, as is already [p. 132] prescribed by order of 19 July 1915 No. 1659/15.A2 for the officers from regimental commander inclusive downward.[130]

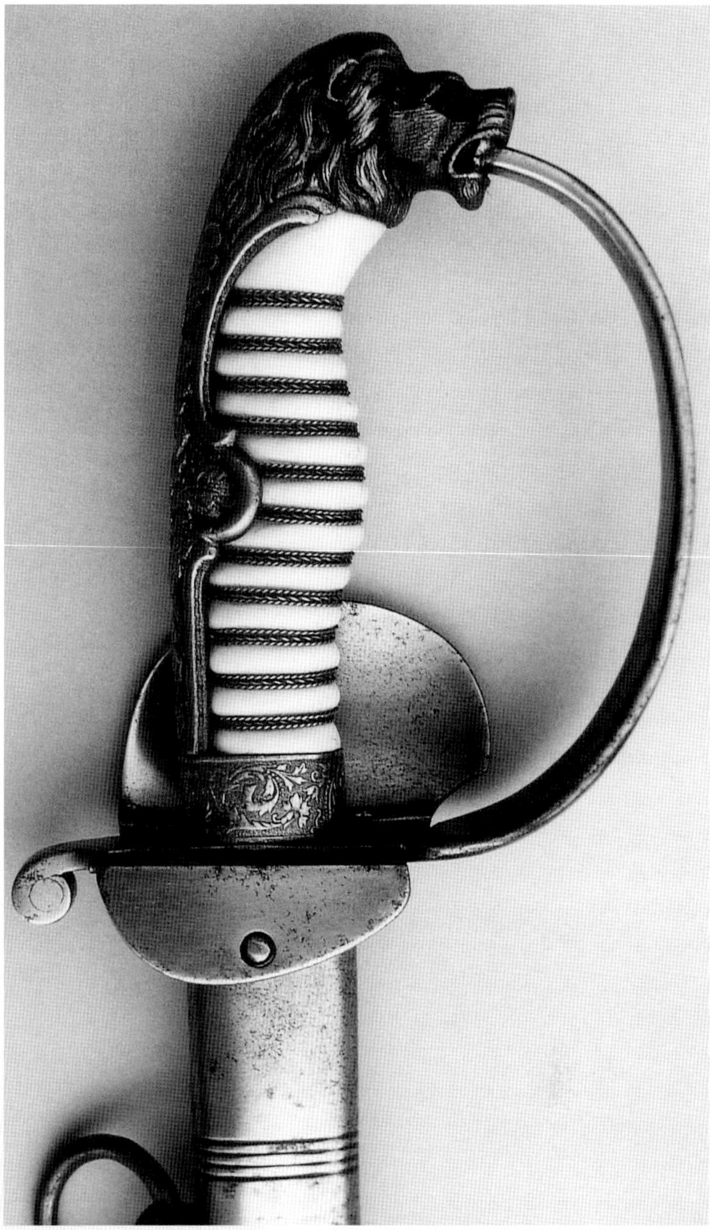

Above: W 177 Hilt of the saber (quarterside) of W 176. (PS)

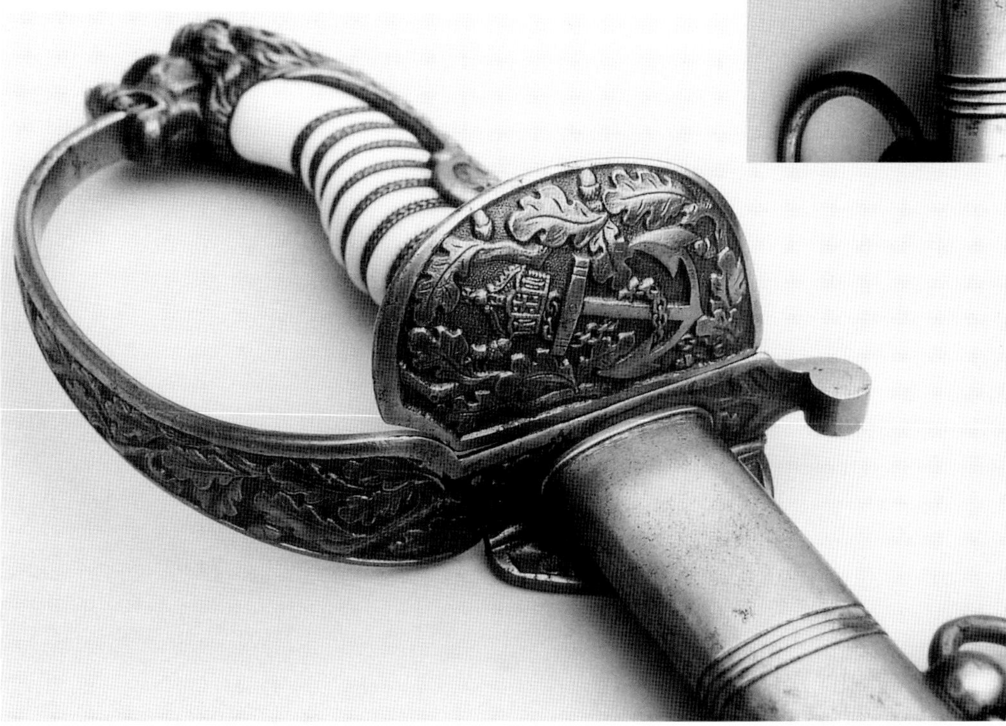

Left: W 176 Basket of a naval saber made of iron (W 175 right). (PS)

W 178 Upper scabbard fitting made of iron. (PS)

W 179 Middle scabbard fitting, as before. (PS)

The view is expressed variously by historians and collectors that the saber no longer came into combat during World War I. It is correct in principle that the sidearm lost significance. Yet there are a few reports that fighting was still done with the saber. One example that a member of the Navy provided shall not go unrecorded.[131]

The officially published Admiralty staff book reports, in the section *The Combat of the Imperial Navy in the German Colonies*—about an advanced observation post during the fighting around the German support point of Tsingtow on the night of November 6-7, 1914—the following:[132]

> *Unfortunately the crew of the advanced post had been surprised by a Japanese engineer troop and cut down except for a vice-sergeant and the telephone operator. The observation officer, Oberleutnant zur See* (Julius, author's note) *Aye, finally fought with his naval saber against Samurai swords. He was found that afternoon, covered with fearful saber wounds.*

W 180 Bottom scabbard fitting, as before. (PS)

The Temporary Reich Navy (1919-1921)

On November 28, 1918, the German Kaiser Wilhelm II abdicated as highest warlord and highest representative of the Imperial Navy. Thus the world's second largest navy[133] lost its promoter, as well as the term "Imperial." As a result of the lost war, personnel was radically limited to 15,000 men. On April 16, 1919, the National Assembly in Weimar passed a law on the formation of the new "Temporary Reich Navy." In a directive from the Reich Defense Minister on August 1, 1919, changes were made in the service uniforms that were stated in the uniform prescriptions for the Navy, uniform code for sea officers, etc.[134] This directive envisioned the sidearm as a "simplified dagger," with no plans for a naval saber.

In the completion of this directive, the details of the "new sidearm" were specified more exactly on November 28, 1919.[135] As the uniform sidearm for the officers and so-called portepee wearers—meaning sergeants and deck officers—a "short dagger" with a black handle and black scabbard was called for. In all, some 3090 persons had the right to wear it.

After April 1, 1920, there could be worn in the Temporary Reich Navy:

1. Sabers of honor of the former Imperial Navy,
2. Weapons inherited from direct descendants, and
3. The Imperial naval saber as well as the Imperial sea officer's dagger after 25 years of service, with war years counted as double.

The Reich Navy still had no naval saber of its own. But the Reich Army, the greater part of the Reichswehr, also had no long sidearm of its own at that time. In a revolutionary and modern era, short sidearms (Sidearm 84/98 and Sidearm 98/05) were prescribed.

Thus ended the days of this "Temporary Reich Navy" in 1921, and there developed, though again reduced in manpower, the actual and now "final" Reich Navy.

46 A photo from the days of the Temporary Reich Navy. Reich President Ebert (2nd from left), along with Vice-Admiral Pu"llen (left) and Admiral Behnke (3rd from left). The two officers wear no swords, but the permitted (Imperial) naval dagger. MSM Archives

The Reich Navy (1921-1935)

On April 5, 1921, the uniforming and equipping of the Reich Navy was codified anew by *Reich President Ebert* with the countersignature of *Reich Defense Minister Dr. Gessler* and *Chief of Naval Command Admiral Behnke.*[136]

As the sidearm for the blue uniform for all ranks from portepee underofficer up, the introduced dagger was specified. All war veterans, as long as they had reached the rank of Feldwebel in the war, were allowed to continue to wear the sidearm they had worn in honor. This regulation also applied to naval officials.

48 Sea officer of the Reich Navy presenting his saber; in the background a portepee underofficer as platoon leader, with his saber hooked on. MSM Archives

With the field gray naval uniform, the short sidearm was specified for all naval members. For so-called small service as well as out of service, all officers, deck officers, and portepee underofficers were allowed to wear the (previous) Imperial naval saber. There was still no new model of a naval saber.

On December 6, 1921, the Reich Defense Minister and the Chief of Naval Command finally issued the *Clothing Guidelines of the Reich Navy.*[137] They regulated, among other things, with what type of uniform the saber was required, allowed, or optional. With all kinds of blue and field gray uniforms, except the tropical uniform, it was now allowed as at least a "permitted deviation." Finally, though, the highest local naval commander was to decide on the composition of the uniform.

47 A portepee underofficer of the Reich Navy with a naval saber. D.S. Archives

49 Field Marshal von Hindenburg reviews a guard of honor. At far right are sea officers with sabers in various positions. MSM Archives

50 Naval officer corps. The music master and portepee underofficer wear naval sabers with their field gray uniforms. D.S. Archives

In 1922 a new officer's saber, the "Uniform Saber 1922," was introduced by the larger service arm, the Army. Now the naval command also took up the idea of its own long sidearm. In the clothing and uniform specifications published by the Naval Commander, Admiral Behnke, it was stated under the heading of *Sidearm for the blue naval uniform:*[138]

a) the dagger . . .

b) the naval saber in a black leather scabbard with brass fittings, after the old (Imperial) pattern. On the basket, in place of the anchor with crown, an unclear anchor appears; the blade is not decorated.

51 A group of Reich Navy men in field gray uniforms, some with naval sabers. The man at right has the I.O.D. 89. Huss Archives

52 Oberleutnant zur See (M. A.) Meisner in a field gray uniform, still bearing his Imperial naval saber. MSM Archives

54 An Oberleutnant of the naval engineers (circa 1935) in dress uniform, with his naval saber attached below. D.S. Archives

W 182 Acceptance stamp of the 1923 naval saber in its first version, made as of 1923: Anchor with letter "M" for "Marine" over it. (PS)

W 184 Acceptance stamp, third version : "M" with stylized eagle above, used from the early thirties to about 1938. (PS)

W 183 Acceptance and naval unit stamps. The acceptance stamp is the second type: "M" with eagle above, the Weimar Republic symbol. The letter and number N 1048 stands for the North Sea Station, weapon number 1048. (PS)

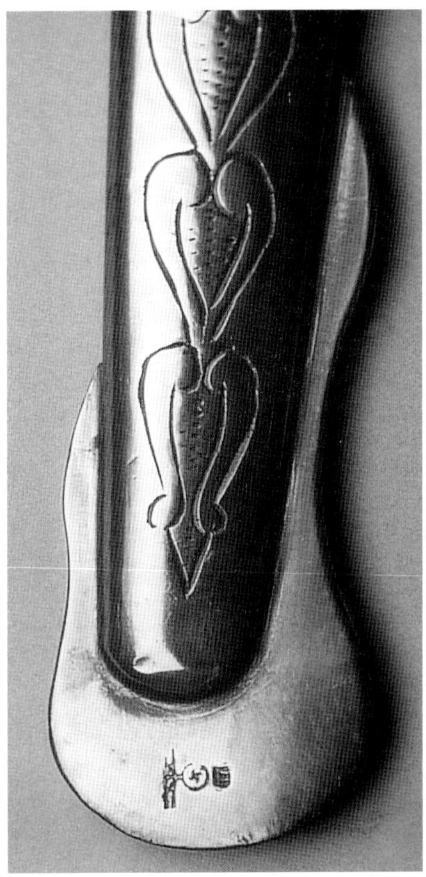

W 185 Acceptance stamp, fourth version: "M" with an eagle with horizintal wings and swastika over it. (PS)

W 186 Naval unit stamp. The letter-number combination O.2607 stands for the Baltic Sea Naval Station, weapon no. 2607. (PS)

And in another place:

All war veterans who wore a portepee are allowed to wear their naval weapon, worn in honor in the war, in an out of service as long as it last....

It remains to be noted that the unclear anchor is 33 mm high and set vertically, without a crown, on the folding outer handguard. This is the essential change to the old Imperial model. The old ordnance sabers were always equipped with fixed baskets.

Description of the service-delivered naval saber of 1923:

The blade is 22 mm wide and is a reinforced type with a brace. On the prescribed service saber, the blade is always undecorated and polished. The trademark of a Solingen manufacturer is customary, but not absolutely necessary.

The lion's head is a finely crafted knob with a high mane, without colored eyes as on the early Imperial naval sabers, the mane running to the grip ring. The actual grip core is made of wood covered with white celluloid plastic with three brass twisted wires

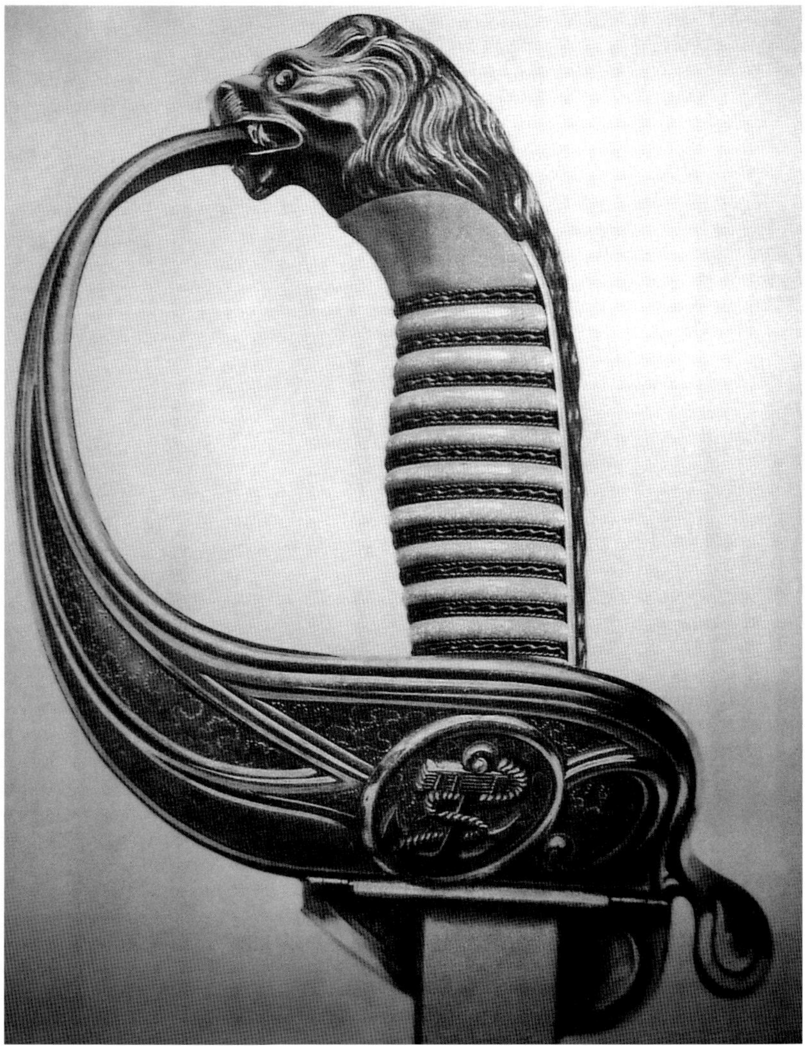

W 187 Naval saber with fixed basket, from a pattern book (circa 1927) of Carl Eickhorn in Solingen. With this saber, which has a fixed basket and an unclear anchor in a medallion, officers who knew the traditions and wanted a saber "like Father wore" were probably being appealed to. Yet few weapons of this type were sold. (DKS)

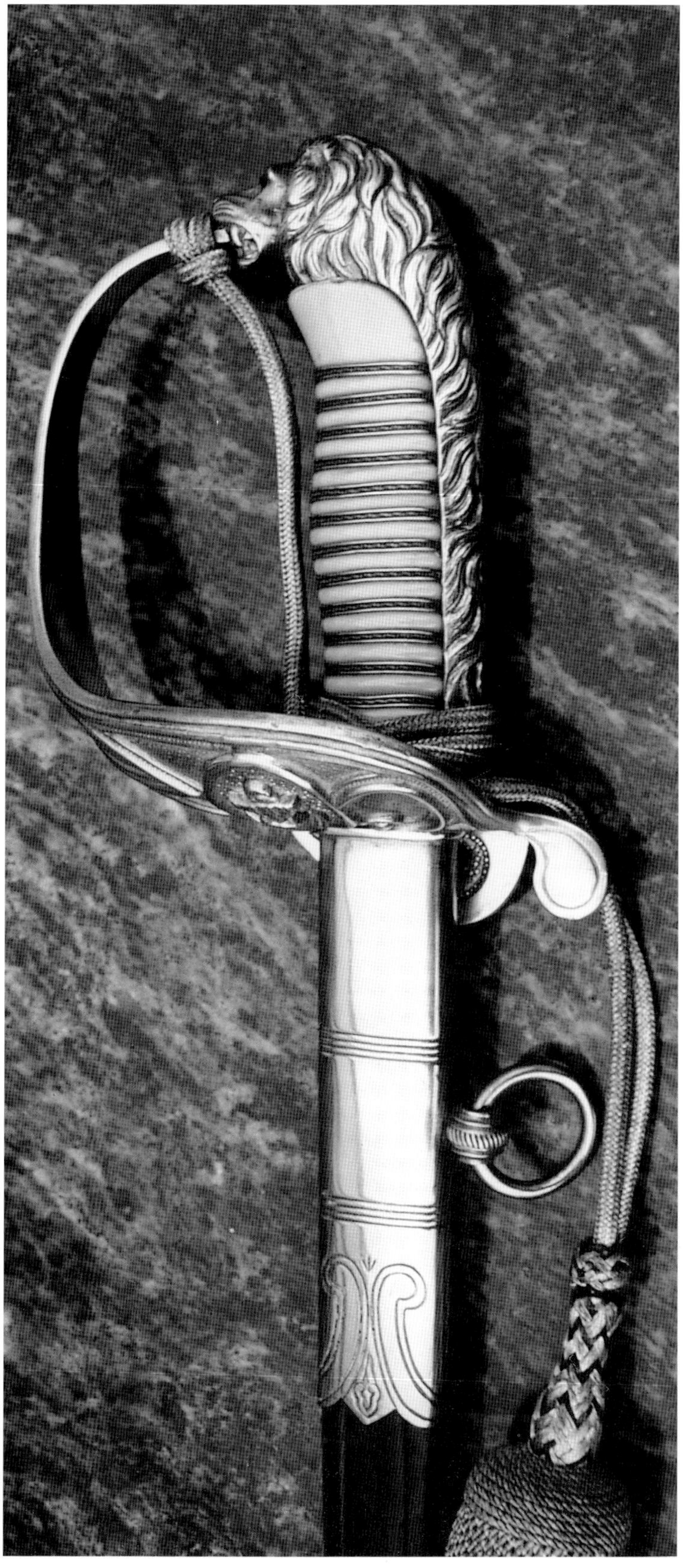

W 188 Hilt of a naval saber with fixed basket. This weapon corresponds to that in the catalog (W 187) but is not made by Carl Eickhorn, rather by Alcoso. The hilt is made of wood and covered with plastic. The portepee is made of spun silver. (PS)

W 189 Fixed basket of a naval saber. The volute to the right of the medallion has a bulge, as do the decorative lines at the left, unlike the 1849 Prussian Navy saber, which has a flat volute. (PS)

wound around horizontally eleven times. Below on the grip is a brass band decorated with oak leaves. The bow of the saber begins in the lion's open mouth and is decorated with tendrils and leaves. On both sides of the guard is a hinge, and the front flap bears the traditional Navy symbol, the unclear anchor. This is based on the emblem of the Prussian Navy from the 1849 era.

Technical Data of examined 1923 sabers (ordnance weapons)				
Weapon	1	2	3	4
Length sheathed	900 mm	900 mm	870 mm	920 mm
Length drawn	884 mm	889 mm	844 mm	885 mm
Blade form	reinforced with brace	reinforced with brace	reinforced with brace	reinforced
Blade length	750 mm	755 mm	710 mm	750 mm
Blade width	22.5 mm	21.5 mm	21.5 mm	21.5 mm
Blade thickness	7.7 mm	7.4 mm	7.9 mm	7.5 mm
Hilt height	130 mm	134 mm	134 mm	135 mm
Grip grooves	11	none	11	
Navy fiscal stamp	M over anchor over M	(Weimar) eagle wings vertical	Eagle over M wings horizontal	NS eagle over M
Weapon stamp	O 179	N 1048	O 949	O 2607

The scabbard consists of stiff black leather with three decorated brass fittings, as was customary on the Imperial naval saber. Two movable rings provide for attachment.

This saber, designated the Naval Officer's Saber Model 1923, was worn by all Navy members from the rank of Feldwebel up from 1923 to 1945. The specimens given out by clothing depots originally had a small clear anchor with the capital letter M as a stamp of acceptance. These weapons, often called chamber pieces, also bear a stamp consisting of the capital letter N (for North Sea Naval Station) or O (Baltic Sea Naval Station), as well as usually three and in rare cases four numerals. Later four-digit numerals were used almost exclusively by the Navy.

The acceptance stamp M with anchor was changed, though, during the Reichsmarine years by order of August 3, 1927.[139] The new emblem was the "Reich eagle" with the letter M. The Reichsmarine used the following types of sidearms:

- Older officers kept their Imperial sabers for reasons of tradition.
- Younger officers carried the new 1923 naval officer's saber.
- Deck officers still wore the Imperial naval officer's saber, as did the older sergeants.
- Younger sergeants and officer candidates from ensign down wore the new 1923 naval officer' saber.

The sabers were issued to sergeants at no cost by the clothing depots. When they left the service, they could buy the saber issued to them for service at a low price.[140] This privilege was withdrawn on November 14, 1935.[141] Officers, who had to supply their own uniforms, had to buy a saber. Thus it was customary that the uniform suppliers for naval officers offered sidearms made by various Solingen firms.

These privately bought weapons, though, did not always conform to the officially accepted pattern. One variant with several special details was very popular. First, the lion's head—as already in Imperial times—had a green starboard and a red port eye, based on the position lights of a ship. The outer movable flap with the large unclear anchor was made somewhat fancier. The anchor was set on an oval plate, which made it somewhat smaller. Oak-leaf decoration was added on both sides. The bow and the grip, minus the lion's head, were also ornamented with oak leaves. There were also models in the Reich Navy that bore laurel instead of the popular oak leaves as decoration. This version, though, was rarely bought and is thus found very rarely today.

All these privately purchased weapons were also supplied with maritime decor on their blades, and there were hollow-ground blades as well as reinforced types. On request, Solingen delivered these private naval sabers with genuine ivory grips. These were reserved, though, for well-to-do officers or portepee wearers. Then, as now, ivory was not cheap. The price lists will give evidence of this.

There were also, as in Imperial times, sabers made with individual dedications. These sabers, known as dedication weapons, must not be confused with the official weapons of honor.

W 191 Basket of a luxurious naval saber. The basket is decorated with oak leaves around the medallion. (PS)

The Kriegsmarine (1935-1945)

With the smooth transition of the Reichsmarine into the Kriegsmarine through the military law of May 21, 1935, nothing changed at first in the area of sidearms. The earlier naval saber of the 1923 pattern remained the officially prescribed naval saber until the surrender in 1945.

Just one change was made in the acceptance stamp for the service sabers that had to be made new, since the upgrading of the Wehrmacht also resulted in a large increase of naval personnel. In place of the "Weimar Reich Eagle" on sabers and scabbards, a somewhat more angular eagle with vertical wings and the capital letter

56 Naval honor guard (7th U-boat Flotilla) in March 1944. The leader of the honor guard still bears the naval saber and belt in wartime. Huss Archives

55 A naval officer (in the war) wearing the naval saber attached below. D.S. Archives

57 A Sea Lieutenant of the Kriegsmarine, carrying a saber with fixed basket. D.S. Archives

M now appeared; its exact date of introduction cannot be determined. This eagle was already replaced formally by order of the Reich War Ministry on July 24, 1936, by the then-prescribed eagle with horizontal wings, officially called the *Hoheitszeichen.*[142] This stamp remained unchanged until 1945. But for reasons of economy, the old stamps were still used for some time, sometimes until 1939. It thus remains to be noted that the service-supplied sabers from the Reich Navy era and those of the Kriegsmarine (1923 to 1945) had the same appearance and can only be told apart by the small acceptance stamp.

In the *Uniform and Dress Directive for the Navy* of 1935, it is stated in the section on the field gray naval uniform that the naval saber could be worn in a black sheet-metal scabbard (known as the steel scabbard today) instead of a leather scabbard. This meant that it "could be done," and that the usual leather scabbard was also allowed with the field gray uniform.

By order of October 7, 1938, the naval saber was to be returned by portepee underofficers when they left the service or were promoted to officer.[143] A saber could be bought back by its former wearer, if he wished, by payment of its assessed value.

The Solingen blade industry, what with the expansion of the Wehrmacht and particularly the Navy, sought customers by offering new models of naval sabers. Striking hilts appeared, such as those with the new national eagle and swastika. This model, made by various firms, found very little interest in the tradition-minded Navy. It is therefore very rare today and not known to many collectors. Naval sabers with fixed baskets, as was formerly customary in the Prussian and Imperial Navies, were also produced and infrequently worn. The anchor had changed somewhat, though, no longer having the massive form seen on the Prussian Navy saber. Overall, the saber appears slimmer than earlier models.

There were also sabers made with new blade etchings popular at that time. Of course, now and then a national eagle with a swastika was shown on the blade. But there are also sabers with blade etchings including motifs of battleships, destroyers, and torpedo boats, as well as amphibian aircraft. All of these blades were made to order and, for an extra fee, also produced with a blue background. One firm (W K C) even produced a blade etching with sailing ships, their flags bearing small swastikas.

For financially better-off persons, naval sabers were made with genuine Damascus blades. Various Solingen firms offered strip Damask, small rose Damask, large rose Damask, Turkish Damask, and the somewhat less expensive Damascus steel. In addition, instead of the usual white plastic handle a genuine ivory handle could be obtained. In present-day naval collections, though, sabers with ivory handles and those with genuine Damascus blades are found very rarely.

Finally, it remains to be reported that at all times sabers were made with etched Damascus blades. These blades were somewhat more expensive than normal blades with decorative etchings, and somewhat less so than the standard Damascus steel blades.

In the war, naval sabers were also worn on suitable occasions by officers and sergeants. This is proved by, among other sources, the many wedding pictures of naval men, who always wore a saber with the so-called grand uniform. On the war front, though, the saber was no longer worn. In 1944 the Wehrmacht High Command forbade officers and Wehrmacht officials in all the fighting forces, including the Navy, to wear a saber, as well as a dagger, by order of December 23, effective immediately. A "sharply loaded" pistol was to be carried on their belt. This practically amounted to retiring sidearms from the Navy.

Weapons of Honor

In the Reichsmarine, as of 1925, the officers, underofficers, and men were awarded prizes of honor according to paragraphs 233 and 315 of the *Shooting Regulations for Rifle, Carbine and light MG* and paragraph 222 of the *Shooting Regulations for the heavy MG.*[144]

W 198 The guard of a 1923 naval saber (W 197). The quarterside form departs from the service model. Here too there is no flap present. This weapon has a blackened (burnished) steel scabbard. In this form it was allowed to be worn by a naval man in a field gray uniform. (PS)

Naval sabers of honor were given to the officers and watches to underofficers and men.

These highly regarded naval sabers of honor had the external form of the 1923 service saber, but their blades were made of Damascus steel. On the blade, in high-relief etched gold lettering, were the name and rank of the marksman, the reason for awarding, and the year of the event. Decorative maritime designs were also etched on.

The "master marksmen" were announced annually in the published naval regulations. A list of all successful winners of naval sabers of honor will be published for the first time below.

This sporting competition of high military importance was carried on annually in the Reich Navy from 1925 to 1935, and the Kriegsmarine carried on the tradition. When the war began in 1939, though, this competition was halted, so that the last prize competition took place in 1938.

In the Reich Navy, only sabers were given as weapons of honor until 1934. In 1935 for the first time a dagger of honor was awarded instead of a saber. Thus the tastes of the times were recognized.

W 194 Naval saber of striking design from the Kriegsmarine era, depicted in the pattern book of the Anton Wingen firm of Solingen (February 1935). A naval saber with fixed basket was offered as number 1859. Sabers 1860 and 1862, with especially designed baskets in Imperial style, with an unclear anchor under a national eagle, surrounded by oak leaves, were sold only rarely in those times. (PS) (Source: Catalog of the Anton Wingen, Jr. firm, Solingen, 1935)

Officers of the Army, Luftwaffe, and Kriegsmarine preferred to wear a dagger. Sabers were worn only for formal occasions. In daily service the shorter, handier dagger was preferred. The traditional saber had now faded into the background. In the Kriegsmarine era, a naval saber of honor was awarded only once.

This exception was made for Oberleutnant (M.A.)[145] Koertge. In the year 1925, Koertge—then a sergeant—had won a watch for being the Navy's best shot, in 1931 he won a certificate of praise, and in 1932 another certificate, still another in 1934, and in 1935 a dagger of honor from the Navy, as well as a silver cigarette case, and in 1935 a dagger of honor from the Navy. Since Koertge had also won in 1936, only a naval saber remained for the Navy command to honor him with. This unique process could never be equaled by any other naval officer. In 1937 the best marksman of all time in the German Navy received a special prize from the Commander of the Kriegsmarine in the form of a silver cup.

It should not be ignored, though, that the conditions for awarding a naval saber of honor were set so high that in the 1928 competitive year not a single naval officer fulfilled them. The criteria for prize marksmanship are given here:[146]

Prizes of Honor

219. In every year, prizes of honor for good marksmanship with the rifle and light machine gun are awarded to officers and underofficers.

a) Number of prizes of honor

220. The number of prizes of honor to be awarded is determined by special order.

b) Shooting with a rifle (carbine)

221. The following set is fired with rifle and carbine:

Distance: 150 meters.

Position: Standing, free hands, target with 24 rings, 5 shots.

At least 100 rings must be achieved.

Outfit: Body belt, cartridge case, sidearm, cap.

Practice shots may be fired before the competition. They must be designated as such in advance.

Primarily, the sum of the rings is decisive. If it is equaled, the winner is he who fired the best last shot, if necessary, the best next-to-last, etc.

222. All officers and underofficers may take part in the competition once every year, if they do not shoot for a prize with the light or heavy machine gun.

223. Allowed to participate are, under existing conditions, the first lieutenants, lieutenants and underofficers of the units of Group A who are at the site on the day of shooting.

c) Shooting with a light machine gun

224. . . . 225. . . . 226. . . .

d) General Rules for Prize Shooting

227. The shooting takes place at the end of the shooting year on as favorable day for the competition as possible. Once the shooting has begun, it is to be interrupted only for urgent reasons—suddenly breaking very heavy rain or storm—and continued on one of the next days.

228. If a soldier who has already won a prize of honor (regardless of whether with the rifle or machine gun) is the victor in a competition in a later year, he is to be noted in the announcement of the results and given a certificate of praise. The next-best marksman receives the prize of honor if he has satisfied the conditions. Thus each soldier can win only one prize of honor.

229. The names of the best marksmen who have fulfilled the conditions are to be published, along with the shooting results, their company etc., and the length of the scabbard (for officers), by October 15 of every year by the defense zone command (etc.). If one of the marksmen has already won the prize of honor, this is to be emphasized. The names of the winners are to be reported to the Reich Defense Ministry (In 2) which approved the issuing of the prizes of honor, by November 1 of that year. The commanders are to award the prizes of honor in a suitable manner. . . .

230. In personal records, in the troop's records, in the shooting records in the shooting book, and in the discharge papers, the awarding of a prize of honor and commendation were to be noted.

List of Officers of the Reichs- and Kriegsmarine to whom weapons of honor (sabers and daggers) were awarded.

1925 Saber of honor	Oberleutnant z. S. Mirus	Cruiser *Nymphe*
1926 Saber of honor	Oberleutnant (Eng.) Wa'Hlisch	1st Unit, Baltic Sea Ship Cadre Division
1927 Saber of honor	Oberleutnant z. S. Oehrl	Naval Intelligence Office South
1828	no saber of honor was awarded	
1929 Saber of honor	Leutnant z. S. Bu"chel	Ship of the Line *Schleswig-Holstein*
	Oberleutnant (Eng.) Kersten	Naval School of Friedrichsort
1930 Saber of Honor	Oberleutnant z. S. Bu"chner	II. Naval Artillery Unit
1931 Saber of Honor	Leutnasnt z. S. von Bogen	Cruiser *Ko"nigsberg*
1932 Saber of honor	Leutnant z. S. Wilcke	Cruiser *Leipzig*
	Leutnant z. S. Opdenhoff	II. Unit, Baltic Sea Ship Cadre Unit
	Kapita"nleutnant Heye	North Sea Ship Cadre Unit
1933 Saber of honor	Leutnant z. S. Foerster	1st Naval Artillery Unit
1934 Saber of honor	Oberleutnant z. S. Gruppe	Blockade Command
	Oberleut. z.S. von Stockhausen	Survey Ship *Meteor*
	Oberleutnant z. S. Rahmlow	III. Naval Artillery Unit
1935 Dagger of honor	Oberleutnant z. S. Lampe	4. Torpedo Boat Flotilla
	Leutnant (M.A.) Koertge	III. Naval Artillery Unit
1936 Saber of honor	Oberleutnant (M.A.) Koertge	V. Naval Artillery Unit
Dagger of honor	Leutnant z. S. Graf	II. Naval Artillery Unit
	Leutnant (M.A.) Krampfer	I. Naval Replacement Unit
1937 Dagger of honor	Oberleutnant z. S. Schuhart	Weddingen U-boat Flotilla
	Fa"hnrich z. S. Lang	Armored Ship *Deutschland*
	Oberleutnant (M.A.) Klar	VI. Naval Artillery Unit
	Oberleutnant (M.A.) Schulz	I. Naval Artillery Unit
	Oberleutnant (M.A.) Kuhn	Fla. K. S.
1938 Dagger of honor	Kapita"nleutnant Bergelt	1st R. Flotilla
	Oberlt. z. S. Kloth-Heydenfeld	Salzwedel U-boat Flotille
	Korvettenkapita"n Bertram	9th Ship Cadre Unit
	Kapita"nleutnant Menge	Torpedo School
	Oberleutnant (M.A.) Schreiber	2nd Ship Cadre Unit
	Oberleutnant z.S. Daublensky von Eichhain	Weddingen U-boat Flotilla
	Fa"hnrich z. S. Kramer	Mu"rwik Naval School
	Oberleutnant (M.A.) Bubert	4th Ship Cadre Unit
	Leutnant z. S, Schu"tze	4th Naval Artillery Unit
	Oberleut. (M.A.) Zimmermann	5th Naval Artillery Unit

Whether sabers of honor were awarded within the Kriegsmarine during World War II could not be determined beyond doubt. It is known, though, that both *Grand Admiral Dr. h. c. Erich Raeder* and his successor, *Grand Admiral Karl Dönitz*, awarded daggers of honor for the Kriegsmarine, as was the official designation.[147] It remains to be noted that in 1938, in peacetime, Grand Admiral Raeder awarded a dagger of honor with diamonds to Admiral Albrecht.

The Project of a new Naval Saber Model

Both the quality and quantity of the Solingen blade industry attained a new high point before the war began. The most varied sidearms were produced for the Wehrmacht, various state officeholders, and officials of the NSDAP. It could not be omitted in this

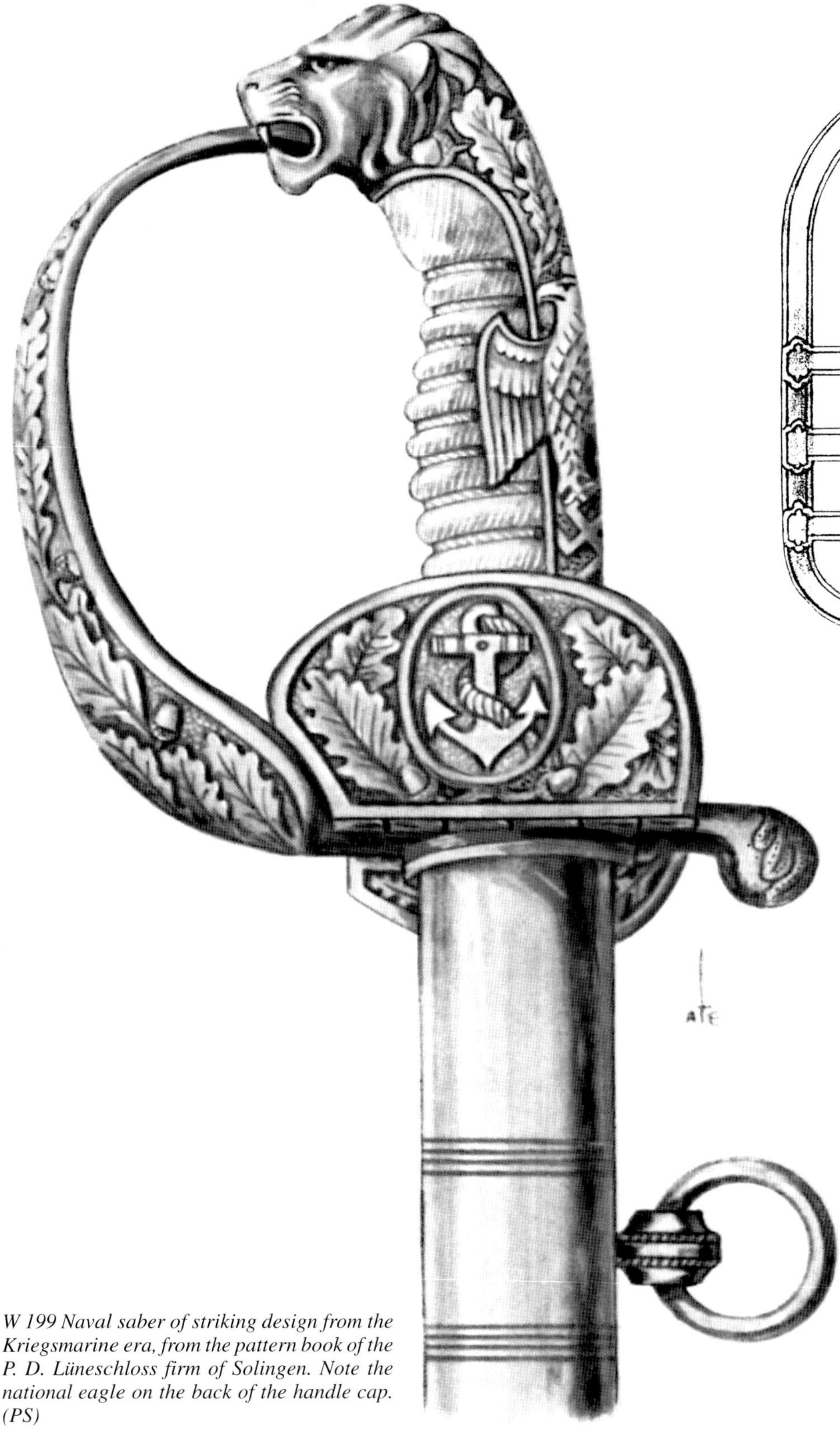

W 199 Naval saber of striking design from the Kriegsmarine era, from the pattern book of the P. D. Lüneschloss firm of Solingen. Note the national eagle on the back of the handle cap. (PS)

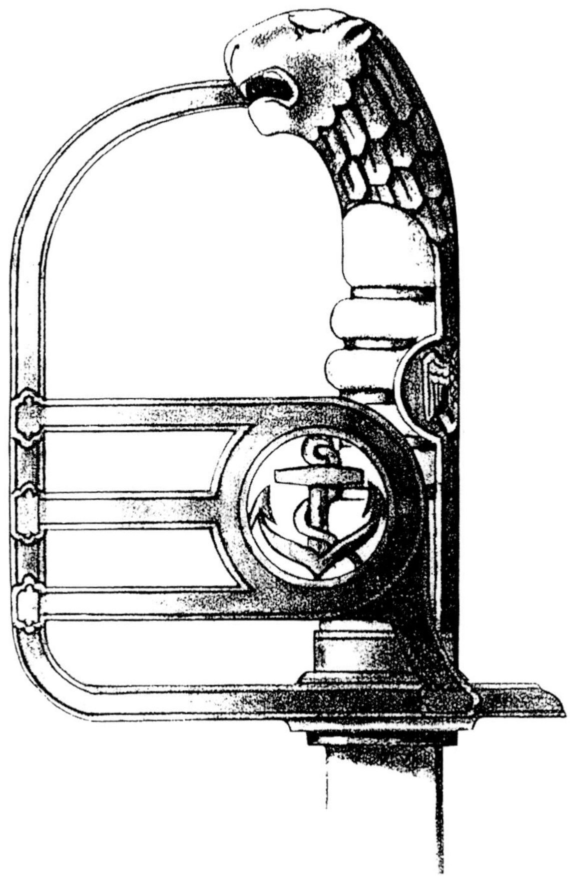

Above: W 200 Design sketch of a naval saber that was probably never made. This study was created by Artrur Eickhorn, then the owner of the Carl Eickhorn firm of Solingen. (PS)

new flowering of sidearms that the old naval saber of 1923 would be replaced by a more modern and timely design. A projected naval saber will be shown here, based on a sketch. The drawing was made by Artur Eickhorn, the son of the well-known Carl Eickhorn, proprietor of the Solingen weapon factory that bore his name. The design is dated 1938.

It is noticeable that, as before, the unclear anchor was the featured emblem on the saber.

The Nazi national eagle was somewhat "hidden" on the handle. The question arises today of whether this was intentional, as the naval officers were always known to be aware of their tradition.

The beginning of the war in September 1939 prevented the introduction of a new sidearm for the Navy. All projected new sidearms, including those of other service arms, had to be halted.

Price List of Naval Sabers, circa 1939	**Reichsmark:**
Naval saber by the Eickhorn firm (no. 1513), with red and green eyes, etched blade and standard celluloid grip	31.35
The same saber, but with genuine ivory grip	42.85
The same saber, but with band Damascus blade	66.35
The same saber, but with small rose Damascus blade	76.35
The same saber, but with large rose Damascus blade	71.35
The same saber, but with Turkish Damascus blade	181.35
The same saber, but with a simple Damascus steel blade	40.35
Naval service saber 1923, as prescribed for sergeants	29.30
Naval saber by ALCOSO, Solingen (no. 308), with red and green eyes, etched blade and standard celluloid grip	31.35
Naval saber by Anton Wingen (no. 1862), with red and green eyes, etched blade and standard celluloid grip	34.00
Naval saber by WKC, Solingen (no. 40), with red and green eyes, etched blade and standard celluloid grip	31.35
Naval saber rack (two-part, with carrying chain)	6.40
Naval saber case (dark blue felt)	1.70
Only for comparison: Walther Pistol PP (7.65 mm caliber), standard weapon of German officers	43.00

Opposite Left: 62 Pilot Captain Paul Wendt (circa 1860), holding a naval saber with black handle. Note the flat-cord portepee. D.S. Archives

Opposite Right: 60 An official of the Waterway Police (first uniform, until 1936) wearing a naval saber attached below. In the background is a naval portepee underofficer. D.S. Archives

The Naval Saber Outside the Navies

Officers, deck officers, and portepee underofficers who retired from the Imperial Navy were, as long as they were not pensioned off for reason of age, utilized further in various civilian institutions that came into contact with the navy and merchant marine. As long as they held state positions, such as harbor officials, fishery inspectors, or civilian pilots, these people were very often allowed to go on wearing their sabers. For example, the regulations for the uniforming of Royal Prussian Fishery Inspectors stated:

Service Uniform.
1. Senior Fishmaster 9and also Fishmaster)
Sidearm: Saber of the naval officers with gold portepee.[148]

In addition, the naval sabers of the Imperial Navy were also worn by army and police officers after World War I, if they were former naval officers, naval officials, deck officers, or sergeants. The Imperial naval saber, though, was then worn in a steel scabbard in the Reich Army manner. A prominent example of this is the later *General of the Paratroopers Bernhard Ramcke* who, as an army officer in the twenties and thirties, always wore his Imperial naval saber in a steel scabbard out of service. At the end of World War I Ramcke was a lieutenant in the naval artillery. Beyond that, at least one high Naval SA leader, *Standartenführer Wilhelm Boltz,* Police President of Hamburg, wore an Imperial naval saber with his blue Naval SA uniform.[149] Until 1919 Boltz was a Kapitänleutnant of the Imperial Navy.

It is scarcely known today that the new naval saber of 1923 was also worn outside the navy. The police officials of the Weimar Republic, and later in the Third Reich too, the members of the blue-uniformed Waterway Police, the State Police Water Guard, the harbor and shipping polices of the states, as well as the coastal police

61 Group photo of high police officers. At the left and right rear are officers with their Imperial naval sabers properly in steel scabbards. The second from left is the Chief of the Ordnungspolizei, General Kurt Daluege. D.S. Archives

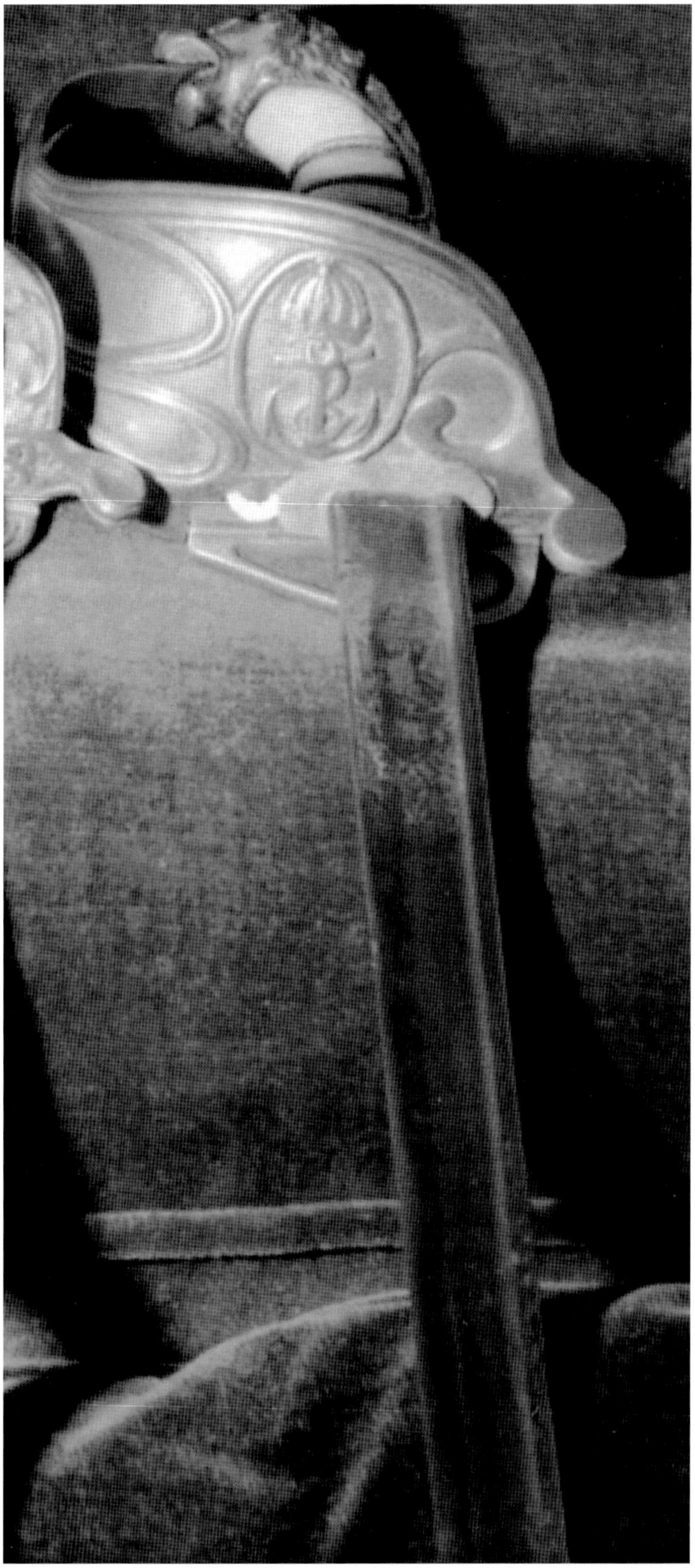

W 201 Basket of a naval saber for Royal Prussian officials in the civilian maritime realm (after 1871). The hilt of the weapon corresponds to the Prussian naval saber of 1849. A difference, though, is the royal crown over the unclear anchor in the medallion. Also, the blade is decorated at its upper end with a Prussian eagle. The handle is white. (PS)

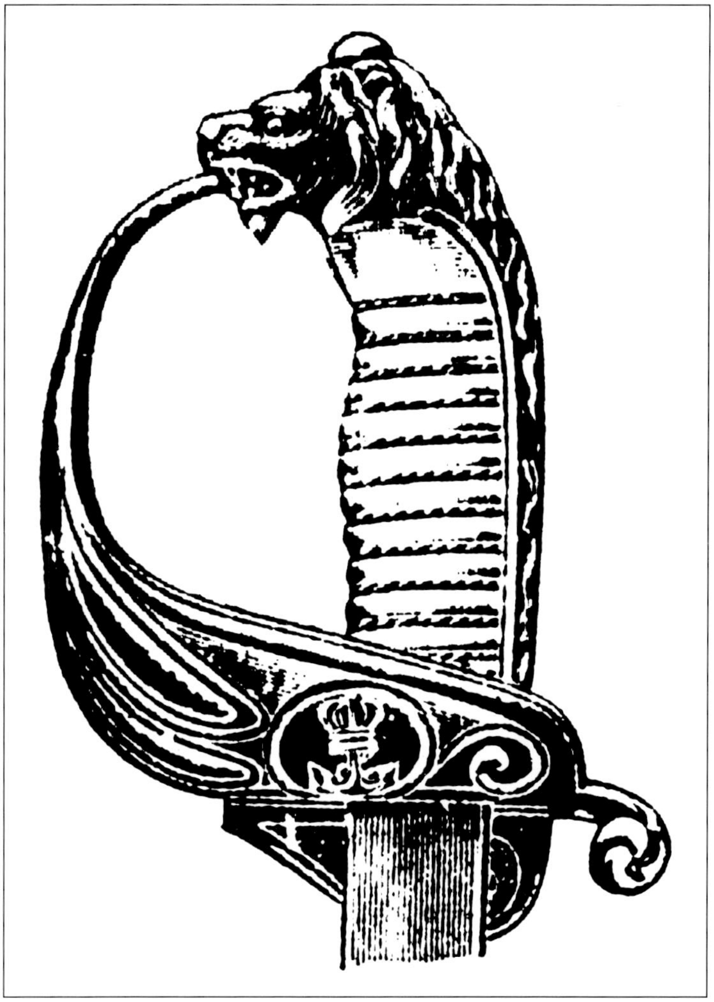

W 202 Naval saber for Royal Prussian harbor police officials. The picture is from the pattern book of the Carl Kaiser firm in Solingen (circa 1911). Here too, there is a royal crown above the unclear anchor. (DKS)

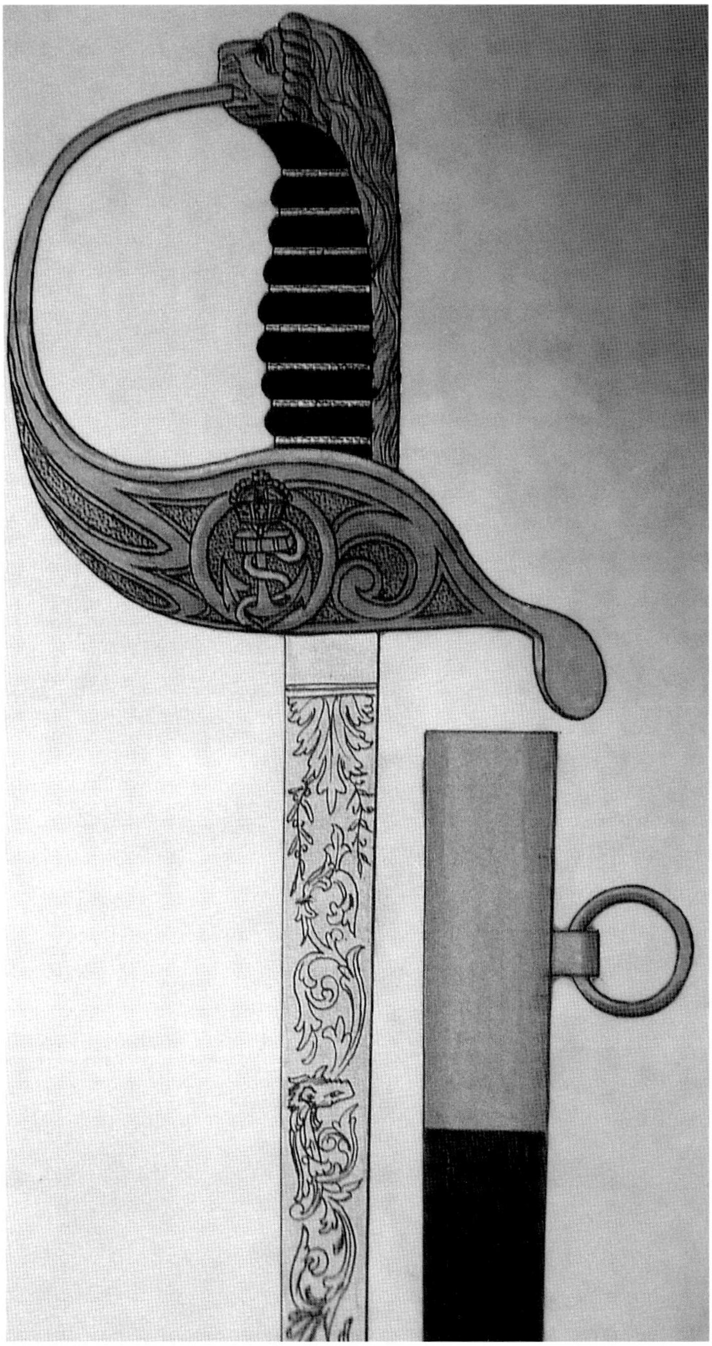

W 204 Naval saber of a pilot from Imperial times. Along with the pierced basket, not allowed in the military sector, this weapon has a black ray-skin grip. (PS)

W 203 Naval saber for pilot commanders and higher-ranking pilots. The picture is from the drawn and colored pattern book—from the early 1870s—of Carl Eickhorn of Solingen. Because of the black grip, this weapon was very probably for a pilot who had a right to wear it, possibly also others in the shipping sector, such as lightship captains, in non-Prussian Germany. For Prussian captains, the Imperial crown was replaced by the royal crown. The scabbard fittings are undecorated. (DKS)

W 206 Basket of the naval saber (W 205) from the post-1923 era. The basket is decorated with oak leaves. The blade and scabbard correspond to the saber for the Kriegsmarine. (PS)

that existed only in the war, wore the naval saber as a long sidearm from the rank of Meister (equal to Feldwebel) up. To be sure, in 1938 a police sword for Führer, and a simpler model for Unterführer, were introduced, but the officials of the Waterway Police were allowed to wear their old (1923 naval saber) sabers. The new police swords were worn mainly by the green-uniformed Ordnungspolizei.

The Oberfischmeister for coastal waters, subordinate to the Reich and Prussian Ministry for Food and Agriculture, and the Fischmeister subordinate to them, also wore the 1923 naval saber with their uniforms, which resembled those of the navy.[150]

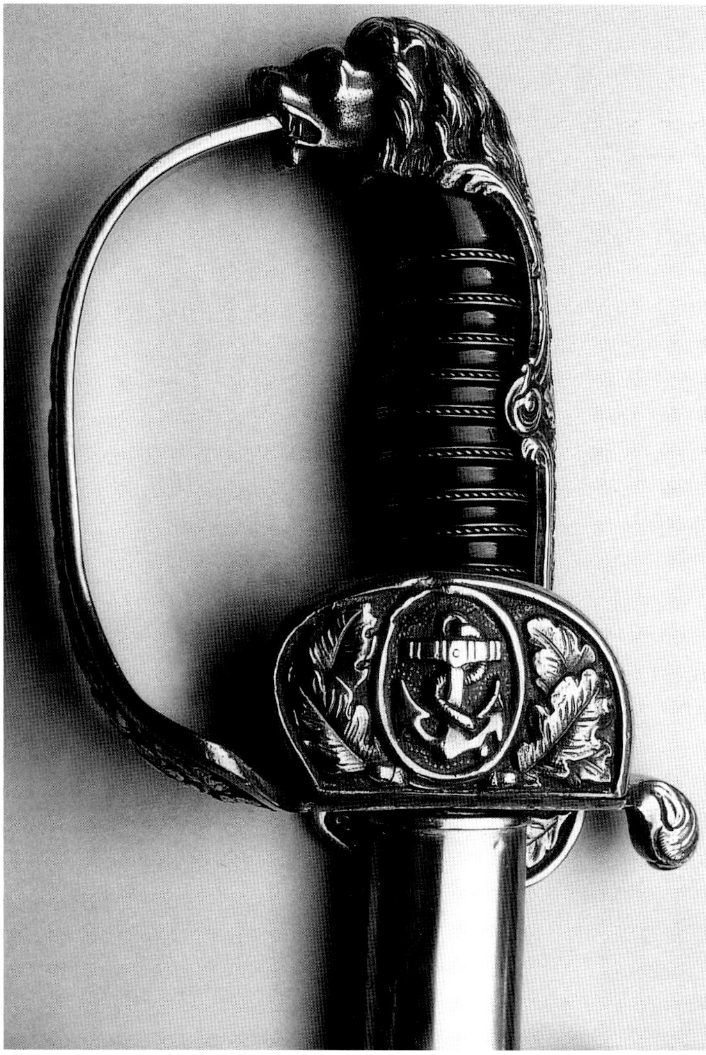

W 205 Hilt of a naval saber from the post-1923 era, made for civilian fishery inspectors, pilots, etc., as the black handle indicates. (PS)

On the other hand, members of the civilian government were sometimes also assigned their own model of saber by the appropriate ministries, but they much resembled the sabers of naval officers.

Sabers for Pilots, Harbor Inspectors, and Harbormasters of Imperial Times

In an order dating from January 26, 1857, a siderarm was prescribed for Royal Prussian Pilot Commanders and for Senior Pilots, as long as they represented the commanders. In the early days it was the 1849 dagger of the Prussian Navy.

After the Royal Prussian Pilot Command was formed from the previous Barsemeisterei[151] on the Jade by order of the Admiralty on April 1, 1862, their uniforms were specified by order of March 3, 1864.[152] For the Pilot Commanders and Senior Pilots, who had the status of military officials, the saber of the sea or deck officers was required. The civilian pilots of these times may also, if they previously wore the Prussian Navy dagger, have been given permission to wear the Prussian Navy saber. After the addition of the Imperial crown to the sabers of sea officers (in 1876), changes were made: They were to have a royal crown on the hilt, and the grip was to be black. In an 1897 book on civil uniforms it is thus stated:

The uniform of the Royal Prussian piloting personnel.

1. Pilot Commanders.

Sidearm: The naval saber, but with the royal crown on the bow and with a polished black horn grip with gold wire winding on a belt to be worn under the service coat, made of black leather with gold fittings with anchor enblems.

2. Senior Pilots.

Sidearm: Only to be worn when representing the pilot commanders is the saber prescribed for them with portepee or a so-called hanger[153] with portepee.[148]

In 1903 the regulations for uniforms of the royal pilots, shipping and harbor police officials, scattered among various offices, were officially collected.[154] The descriptions of the sabers were identical to the former ones and do not need to be repeated. The same saber was required for harbor inspectors and harbor-masters.

Other Naval Sabers

It should not go unnoted that in Imperial times, in the Weimar Republic and even the Third Reich, various manufacturers turned out small naval sabers, some of them of the highest precision, as toys for children. These not exactly cheap toys are now collectors' items for the "big boys."

Naval Sabers After 1945

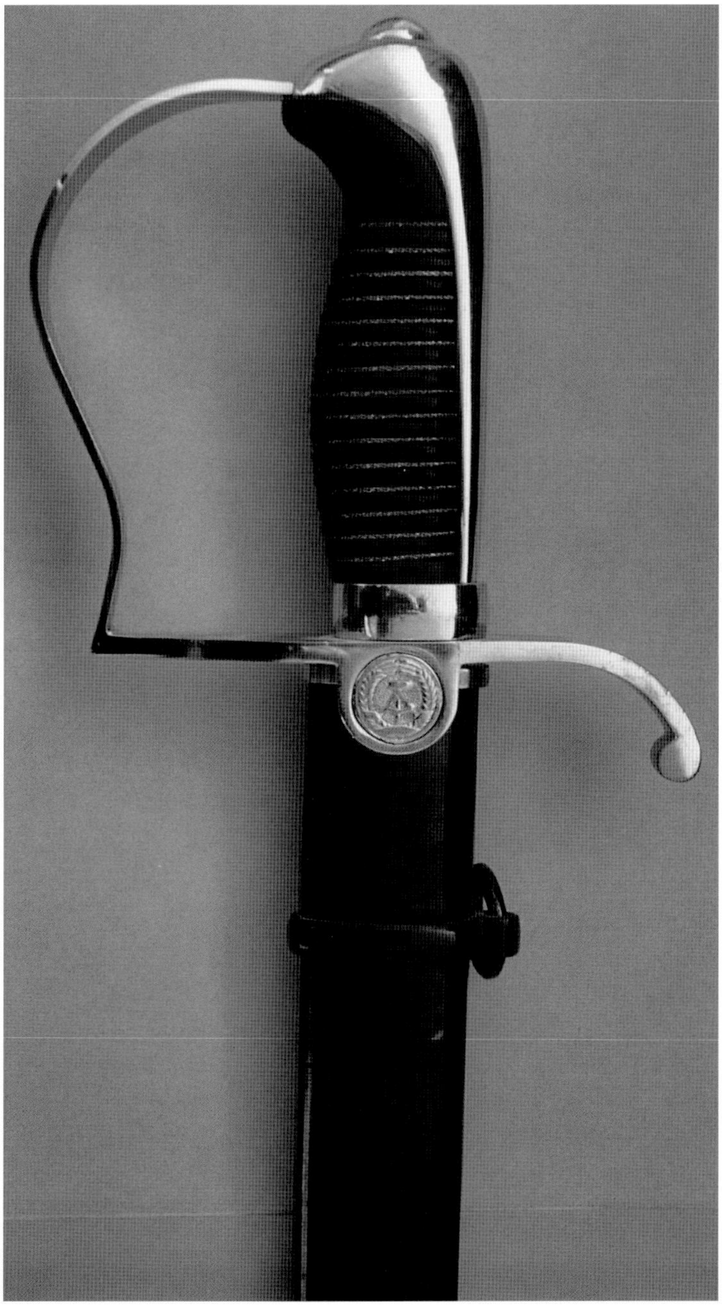

207 NVA officer's saber. (SPT)

On May 7, 1945, the German Wehrmacht surrendered. The four victorious powers ordered that all kinds of firearms and sidearms be turned in. The possession of weapons of any kind, including naval sabers, was now forbidden, and those who disobeyed were imprisoned.

But to some extent, especially by the British naval authorities, the former servicemen were allowed to keep sabers and daggers if the blades were shortened by some 5 to 10 centimeters. To this day, such weapons with cut-off blades are found in many collections.

For some years there seemed to be no more need for sidearms, including naval sabers. But in the early fifties, German Navy sabers began to be produced again in Solingen. In part, spare parts still on hand from prewar production were used in these sabers. These "new naval sabers" were offered, for example, in the MOH News[155] in 1953 for 69 DM. It is documented that, among others, the firms of E. & F. Hörster and Carl Eickhorn were engaged in this production.

Also speaking for the popularity of the German naval saber is the fact that copies of the weapons were made in the eighties in a "cheap labor country" in Asia. These "replicas" were marketed all over Germany by, among others, a large South German hunting supply house. The quality of these products differs markedly from that of the postwar Solingen products.

The members of the Bundesmarine have never officially worn a saber. Unofficially, though, if one believes the tales of old "sea dogs," there have been naval officers here and there who have been seen with a saber, especially in South America.

In the East it has been quite different. The People's Navy of the former NVA used both sabers and daggers. The Volksmarine saber, though, was not a special naval saber, but a uniform saber for all service arms of the NVA. Therefore it is not in the direct tradition of the German naval saber.

The Portepee

Since the beginning of the 18th century,[156] the so-called field emblem that was to be tied around the hilt of the weapon always went with the officer's sidearm. The term "field emblem" was replaced in the course of time by the French term "porte-epee," which basically means "sword-carrier." The portepee, which gave special honor to the wearer, served as the officer's symbol of rank.

In the course of time the higher officials, such as the postal and mining officials, were given the right to have a portepee. Unlike the officers, though, this was not silver, but gold, and was also called a "Degenquaste" (sword-tassel) at first. In Prussia, though, the silver one was regarded as more valuable. Officers who left military service were not allowed to wear the silver portepee after 1795 unless they had the express permission of the King.

W 209 Portepees (from W 208), with dark blue filling of the "acorn" at left and black-white-red at right. (PS)

Since March 21, 1789, all older underofficers in Prussia were allowed to wear this officer's symbol "for honor's sake." The well-known concept of "underofficer with portepee" of portepee underofficer derived from that. But this was not true in all German lands; the Kingdom of Bavaria followed this example only in 1873.

When a portepee is mentioned in connection with naval officers before 1849, it thus refers to either the "civil portepee" or the portepee that infantry officers usually wore. So according to an order of July 12, 1816, *the uniform cut like the uniforms for the infantry officers* was required for the officers taken into Prussian service from Sweden after the Wars of Liberation.[157] The sea officers' portepee thus could correspond to that of the infantry officers. Until the beginning of the 1820s, the silver portepee also had loose tassels.[158]

With the reorganization of the Prussian Navy in 1848-49, the Navy took over the pattern of the infantry officers' portepee, but with the difference that a round cord was required instead of the flat cord.[159] It was matte silver with black silk filaments in the cord, slide, and stem.

Obviously, the change was based on the pattern of the British portepee. In Britain the round cord was drawn through two round holes in the guard and finally attached to an oval hole in the bow. This type of attachment never worked out for the Prussian naval portepee, since the round cord would have had to be divided for it.

W 208 Portepees. At right is the silver type with black and red for officers, portepee underofficers, etc. At left is a special version of the gold type with dark blue threads for naval officials. (PS)

Those who had the right to wear this type of portepee were all sea officers, sea sergeants, deck officers, and underofficers with portepee.

The Prussian naval officials, by order of May 2, 1856, wore a silver portepee of the same type, but with blue silk threads instead of black running through it.[160] A gold or spun gold portepee with blue silk threads, though, was allowed for one-year volunteer doctors.

The officers and sergeants of the naval infantry, as long as they wore the fusilier officers' saber, could wear the infantry officers' portepee. With the introduction of the naval saber, they wore the sea officers' portepee with it. The naval artillery officers also wore the infantry portepee at first, then as of about 1868 that of the Prussian cavalry.

For the officers, deck officers, doctors, and administrative personnel of the Reich Fleet of 1848, a portepee in the national colors, still referred to as a Degenquaste, was prescribed in the clothing requirements of March 1, 1849. The national colors were black, red and gold.

For the officers and other portepee wearers of the Scheswig-Holstein Navy, there was obviously also a portepee of their own. To be sure, the historical sources regarding uniforms are lost,[161] but in a museum there exists a well-protected portepee that must be so classified according to its colors. The colors blue, white, and red correspond to those of the Schleswig-Holstein cockade of the time.

After the Imperial Navy was founded by order of February 24, 1874, new uniform regulations for officers, doctors, military officials, etc. of the Imperial Navy were issued, and were corrected and expanded in the following years. Of the new portepee for sea officers, which matched the Imperial colors of black, white, and red and replaced the silver Prussian type with black threads, it was stated:[162]

> *14. Portepee. Of silver worked with black and red silk, on two round silver cords worked with black and red, which are united.*
>
> *For the officers and portepee wearers of the ses artillery it was said further:*
>
> *10. Portepee. Of silver worked with black and red silk, on a black leather strap, 1.8 cm wide, plaited with silver and red.*

Who wore which portepee in Imperial times can be seen in the following list:

A. Portepee with cord, slide, and stem in silver with black and red silk threads, filling of the ball in black-white-red:

- I. Sea officers
- II. Engineer officers
  - 1. Naval engineer officers
  - 2. Torpedo engineer officers
  - 3. Firework, ordnance and torpedo officers
- III. Medical officers
- IV. Naval paymasters
- V. Deck officers
- VI. Feldwebel including Vice-Feldwebel
- VII. Naval infantry officers
- VIII. Naval musical directors
- IX. Higher personnel for ship- and machine-building (officials)

B. Portepee with cord, slide, and stem in silver with blue silk threads, filling of the ball in blue:

- I. Naval justice personnel
- II. Naval quartermaster personnel
- III. Piloting personnel
- IV. Naval pharmacists
- V. Weaponmasters (after 15 years' service)

C. Portepee with cord, slide, and stem in gold with blue silk threads, filling of the ball in blue:

- I. Garrison-building personnel
- II. Clothing and provisioning office personnel
- III. Higher harbor-building personnel.
- IV. Shipyard company personnel
- V. Naval lazaret personnel
- VI. Garrison administrative personnel
- VII. Shipyard administrative personnel

With the establishment of the Reich Navy, the uniform regulations were changed step by step. In 1920 (on December 27) the Admiralty worked out a plan for the uniforming and equipping of the blue-clad naval units, according to which the portepee was to be *with a completely silver portepee cord (without black and red threads).* With the issuing of the final *Regulations for the uniforming and equipping of the Reichsmarine (blue and field gray)* by Reich President Ebert on April 5, 1921, the new dagger portepee was finally introduced,[163] that for the saber remaining—temporarily—the old Imperial portepee. Only by order of the Naval Commander on May 8, 1923, was a new model of the portepee for the naval saber introduced—along with the new model of the naval saber. The portepee for the saber was to be the former saber portepee, but with a completely silver cord (without black and red threads).[164]

In the Kriegsmarine era too, the portepee remained unchanged in form. There exist, though, both silver peacetime portepees and those made of aluminum thread. There was one difference in price. The "genuine silver" portepees were considerably more expensive. In World War II some were also made of artificial silk, an "ersatz material." These portepees change their color under strong light. They become yellowish and look gold-colored today. But it should be noted that no gold portepees were worn in the Reichs- or Kriegsmarine.

Dimensions of Saber Portepees			
	Prussian Navy	**Imperial Navy**	**Reichs-/Kriegsmarine**
Length of ball without stem	46.0 mm	53.0 mm	46.0 mm
Diameter of the ball (max.)	32.0 mm	40.0 mm	33.0 mm
Length of the stem with slide	46.0 mm	40.0 mm	40.0 mm
Length of the cord	505.0 mm	480.0 mm	500.0 mm
Thickness of the cord	5.0 mm	5.0 mm	4.5 mm

Abbildungen der schönsten Muster von echtem Damaszener -Stahl.

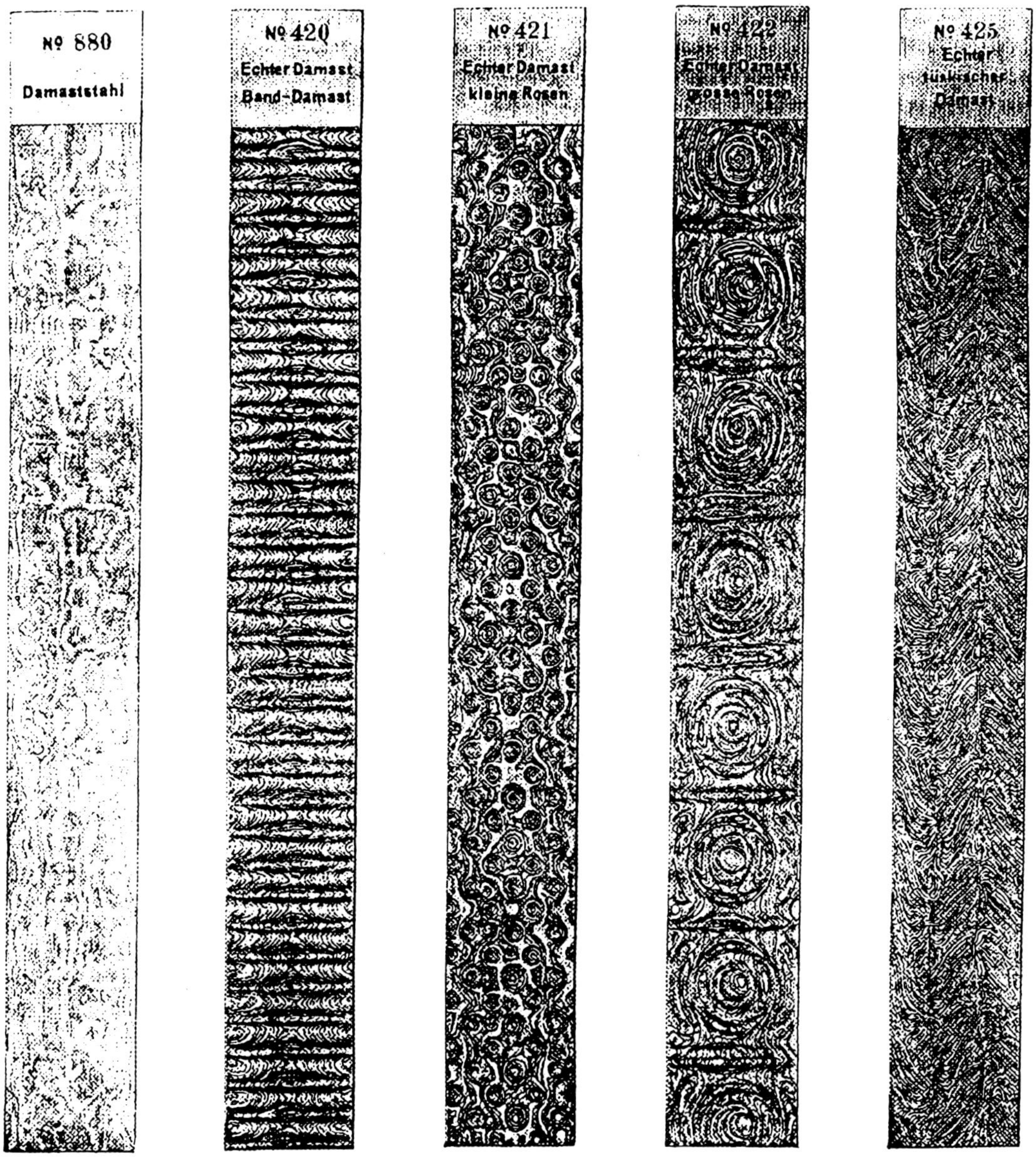

Die Vorzüge des echten Damaszener Stahls bestehen in seiner grossen Widerstandsfähigkeit.

Die sehr alte Herstellung des Damaszener Stahls soll durch Not entstanden sein, indem man aus Mangel an Stahl alte Eisenstücke zusammenschweißte. Alte Dolche und Schwerter deuten darauf hin, daß der Ursprung der Erfindung Nordindien ist.

Durch die Förderung der Waffenfabrikation in Damaskus entwickelte sich die Kunst der Herstellung von Damaszener Stahl zur höchsten Vollkommenheit.

Der Damaszener Stahl entsteht durch inniges Verschweißen von harten und weichen Stahlarten und durch wiederholtes Ausschmieden des dadurch erhaltenen Stahls. Die Zeichnungen im Damaszener Stahl werden durch Beizen erzeugt, nachdem die Oberfläche glatt geschliffen ist. Die in der Säure enthaltene Beize färbt die Fasern der verschiedenen Stahlarten hell und dunkel.

Die Zeichnungen im Damaszener Stahl. als: kleine und große Rosen, Mosaik pp, sind auf das Nebeneinanderlegen verschiedener Stahlstäbe in der gewünschten Form zurückzuführen. Durch die ungleiche Härte u. Porösität der verwandten Stahlarten entsteht beim beizen die geschmiedete Zeichnung.

Beim echten Damaszener Stahl kann man die Zeichnung abschleifen und dann durch Beizen immer wieder hervorrufen. Dieses unterscheidet den echten Damaszener Stahl von Nachahmungen.

Der höchst beträchtliche Abbrand bei den vielen Schweißungen bedingt einen großen Materialaufwand, hierdurch und durch den großen Zeitaufwand, welchen das Schmieden der Zeichnungen erfordert, erklärt sich der hohe Preis des echten Damaszener Stahls.

Document Titled: Illustrations of the nicest models of genuine Damascus steel Source: Catalog of the August Lüneburg firm, Flensburg, no date.

Appendix

Glossary

Q 1:	R. Brommy & H. von Littrow, Die Marine, Leipzig 1878.
Q 2:	Transfeldt et al., Wort und Brauchtum im deutschen Heer, Hamburg 1976.
Q 3:	G. Seifert, Fachwörterbuch der Blankwaffenkunde, Haiger 1981.
A.K.O.:	Direct order from the King of Prussia and later the German Emperor.
Anchor, unclear:	Anchor (stock, normal or Admiralty anchor) with a cable or chain around it.
Arabesques:	Decorations formed of flowers, twigs, etc.
Bandoleer:	Shoulder belt for swords, sabers and other sidearms; Q 3, p. 4.
Barge:	Flat-bottomed craft with one mast, pointed ends, and swords on the sides.
Brace:	See drawing of saber.
Cadets:	Students in military and maritime training institutions.
Carp-tongue:	Blade with central point at end and rhomboid or lens-shaped cross-section.
Cheeks, blade:	See drawing of saber.
Core:	Actual grip of a saber, covered with leather, fish skin, etc.
Cuffs:	Turned-up sleeves, which may be decorated with braid; Q 2, p. 166.
Damascus:	High-grade steel; see the graphic on p. 164.
Dull edge	Upper part of a saber blade without a sharp edge.
Effects:	Totality of all the emblems on a uniform.
Epaulettes:	A last remnant of knighthy attire that developed over the shoulders of the armor, originally for protecting shoulders against sword cuts. Officers received them later as a sign of rank; Q 2, p. 62.
Espingen	Small ships.
Facings:	Lapels of a coat descending from the collar, often decorated; eliminated in the Prussian Army after 1809; Q 2, p. 165f.

Gangway: Ramp from a ship to a pier.
Grip cap: See drawing of saber.
Installation: Document of appointment to an office.
Naval Dept.: Naval administrative office.
Pandour point: Special type of tip with back blending into point concavely.
Point: See drawing of saber; tip of the blade.
Quartside: Side of a weapon turned toward the wearer.
Rifle repair fund: Money issued to a troop for weapon repairs.
Sheathed: What a sidearm is when it is in its scabbard.
Terzside: Side of a saber facing away from its wearer's body.
Test model: Specimen of a weapon, uniform or equipment piece exactly conforming to the model and used as a pattern for acceptance; marked and sealed as such.
Trailing saber: Saber that is not worn straight and (actually or almost) drags on the ground.

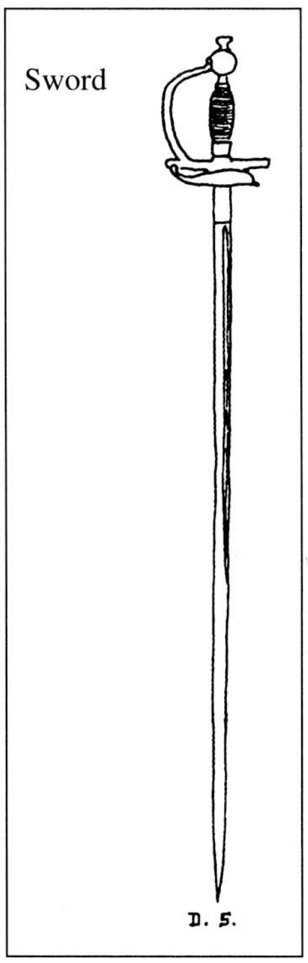

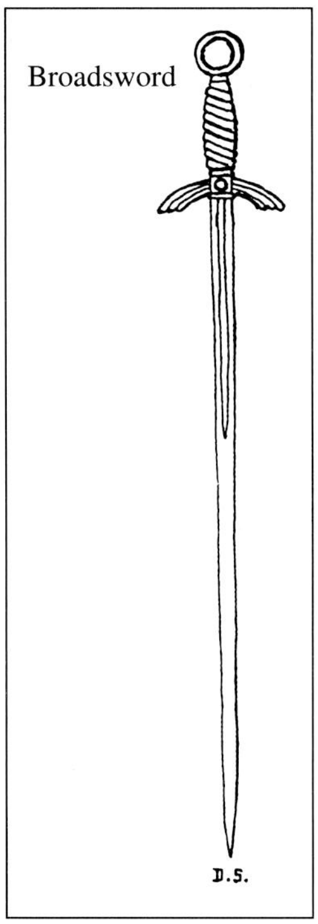

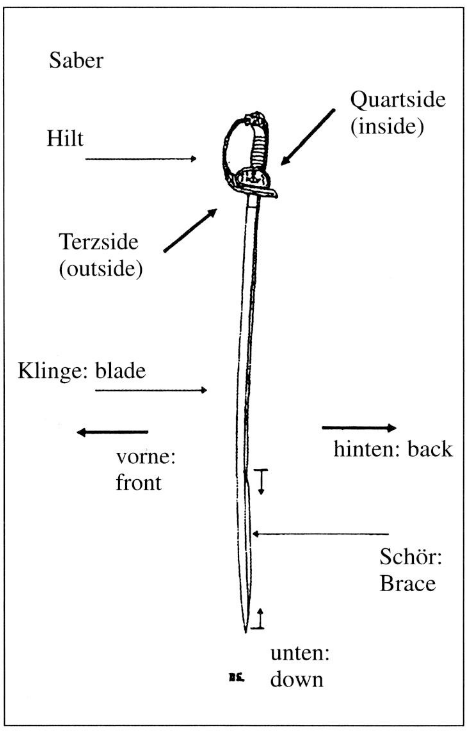

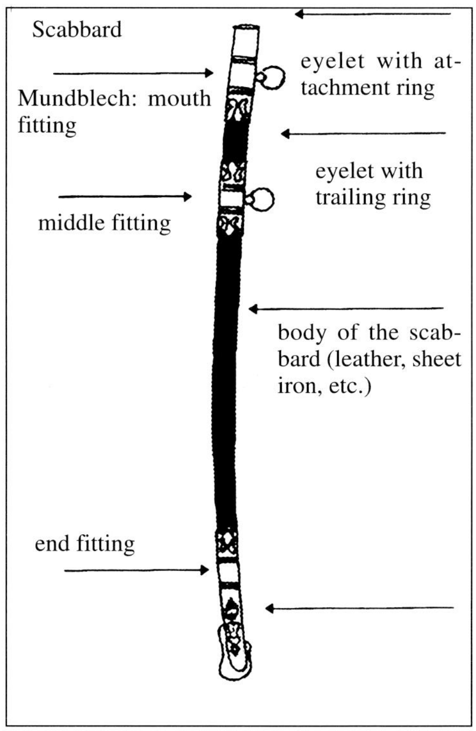

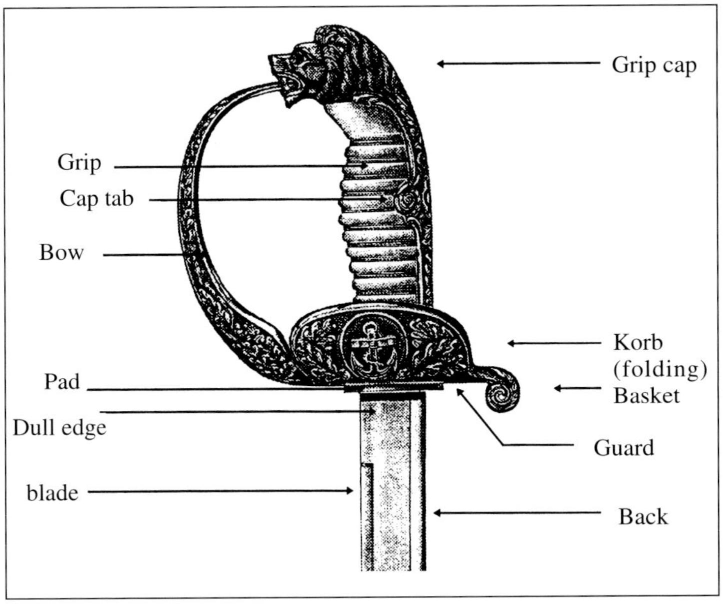

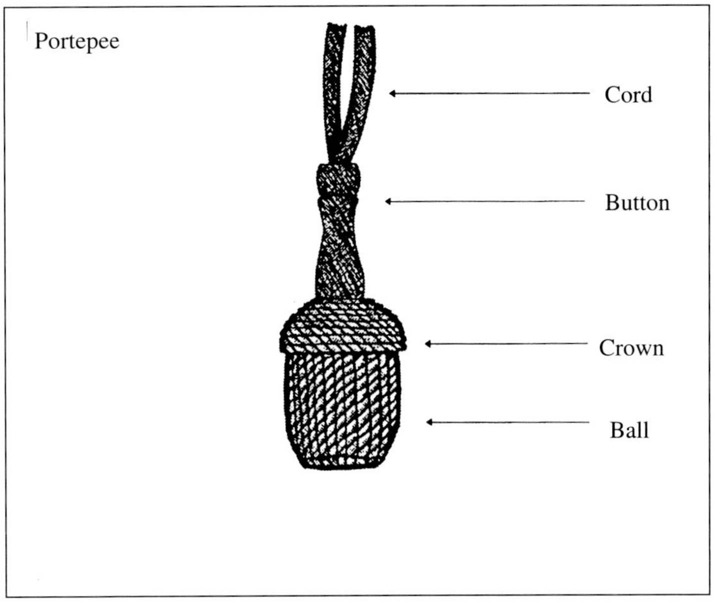

AUG. LÜNEBURG
Kiel, Dänische Straße 22 : Flensburg, Holm 35
Lieferant der Offizier- und Seekadetten-Kleiderkassen

Säbelkoppel

	Mk.	Pf.
1 Säbelkoppel die Beschläge des Gurtes neu vergolden oder versilbern	2	75
2 Paar Karabinerhaken und die Kette der Lederstander vergolden oder versilbern	2	—
1 neuer Säbelkoppelgurt ohne Beschlag .	4	—
1 neuer Säbelkoppelgurt mit Beschlag . .	7	75
1 Paar neue geflochtene Lederstander mit Kette	4	75
1 neues Säbelkoppel aus bestem Moireeband mit doppelt vergoldeten Beschlägen und Patentschieber	12	50
1 neues Säbelportepee Ia.	2	50
1 neues Säbelportepee IIa.	2	—

Zur Aushilfe stehen bis zur Fertigstellung von Reparaturen gute Säbelkoppel zur Verfügung

AUG. LÜNEBURG
Kiel, Dänische Straße 22 : Flensburg, Holm 35
Lieferant der Offizier- und Seekadetten-Kleiderkassen

Marinesäbel

	Mk.	Pf.
1 Marinesäbel vergolden, Klinge und Griff aufpolieren, die Lederscheide auffrischen	11	50
1 Marinesäbel neue Lederscheide einziehen 3.— bis	3	50
1 Marinesäbel alle Teile aufpolieren und auffrischen der Scheide	2	50
1 Marinesäbel das Ortband der Scheide mit neuem Schlepper versehen	1	50
1 Marinesäbel kürzer machen.	2	50
1 Marinesäbel die Klinge scharf schleifen	1	—
1 Marinesäbel neuen echten Elfenbeingriff je nach Größe . . . . 9.— bis	12	—
1 Marinesäbel neuen imit. Elfenbeingriff .	3	50
1 Marinesäbel neue Klinge . . . 5.— bis	10	—
1 neue Marine-Säbelscheide, vergoldet. .	12	50
1 neue Marine-Säbelscheide, nicht vergoldet.	9	—
1 neues Marine-Offizier-Säbelportepee Ia.	2	50

In neuen Marine-Säbeln halte ich stets viele schöne Muster vorrätig, Preise billigst, Preislisten mit Abbildungen kostenlos zur Verfügung

Zur Aushilfe stehen bis zur Fertigstellung von Reparaturen gute Säbel gerne zur Verfügung

Two pages from the Lüneburg firm's catalog, offering parts and repairs for sabers and accessories.

Notes

1 Josef Zienert, Unsere Marineuniform, Hamburg 1970, p. 7.
2 Curt Jany, Das Preussische Heer, Vol. II, Osnabrück 1967m p. 612.
3 Max Schmitt-Hartlieb (ed.): Joachim Nettelbeck—Bürger zu Colberg, Leipzig-Berlin 1909, p. 60.
4 Hermann Franke, Handbuch der neuzeitlichen Wehrwissenschaften, Vol. II, Das Heer, Berlin-Leipzig 1937, p. 518.
5 Hans Szymanski, Brandenburg-Preussen zur See 1605-1815, Leipzig 1939.
6 Paul Koch, Beiträge zur Geschichte unserer Marine, Berlin 1900, p. 69.
7 Zienert ascribes an appropriate sword or short dagger to naval officers of that time; see page 9.
8 Already in 1836, as Crown Prince, he had spoken with his cousin Adalbert about the formation of a navy.
9 Eberhard von Manthey, Deutsche Marinegeschichte, Charlottenburg 1926, p. 45.
10 Zienert, op. cit., p. 21f.
11 Jan Pieter Puype, Der Marinedolch, Deutsches Waffen-Journal, No. 3, 1974, p. 292f.
12 The Prussian marine dagger is said to have been 17 to 19 inches, or 44.5 to 49.7 mm long.
13 Walter Hubatsch, Die erste deutsche Flotte 1848-1853, Herford 1981, p. 89.
14 By "sword-tassel" the portepee is meant.
15 Jörg Duppler, Der Juniorpartner—England und die Entwicklung der Deutschen Marine 1848-1890, Herford 1985, p. 47f.
16 This comes from a document of the Naval High Command, dated August 31, 1849, to the Royal Naval Office in the War Ministry.
17 Adalbert Mila, Uniformierungsliste der Königlich Preussischen Armee und Marine für das Jahr 1869, 1st ed., Berlin 1869, p. 87.
18 Mila, discussing the officers' weapons, speaks particularly of a brass hilt, mentioning it again in the second edition (1872) of his work; he mentions a gilded hilt only in the Imperial Navy.
19 Robert Wilkinson-Latham, Degen und Schwerter in Farbe, Munich 1978, p. 27.
20 Rudolf Cronau, Geschichte der Solinger Klingenindustrie, Stuttgart 1885, p. 18 & Table 2.
21 John Walter, The Sword Makers of Imperial Germany 1871-1918, Brighton 1973.

22 H. von Holleben, Deutsches Flottenbuch. Erlebnisse eines See-Kadetten in Krieg und Frieden, Leipzig 1892, p. 150; author's boldface.
23 John Walter, op. cit., see information on both firms.
24 von auxiliarius (Latin): "makeshift-"
25 The Cadet First Class corresponds to the later Leutnant zur See. [p. 169]
26 Sea cadets here correspond to the later Fähnrich.
27 Allgemeiner Marine-Befehl (General Naval Order) No. 152, p. 126.
28 MWBl 1849, p. 95f.
29 MWBl 1850, p. 273.
30 Allgemeiner Marine-Befehl No. 31, p. 33.
31 Allgemeiner Marine-Befehl No. 83, p. 213.
32 Adalbert Mila, op. cit., p. 87.
33 Allgemeiner Marine-Befehl No. 48 of June 1, 1858, p. 145f.
34 MWBl 1850, p. 54f.
35 Collection of the Service Regulations of the Naval Guard Corps, Berlin 1839, p. 161.
36 Wolfgang Jahnke, Beiträge zur Geschichte der Marinebekleidung, Marine-Rundschau 1941, p. 159.
37 AVBl 1881, p. 261, No. 247.
38 Hettler, See-Artillerie Abteilung, Matrosen-Artillerieabteilungen, Zeitschrift für Heereskunde 1933, p. 581f.
39 See more on this subject in, among others, Hubatsch, p. 29ff.
40 Jörg Duppler, Prinz Adalbert von Preussen, Herford 1986, Appendix 1, complete text.
41 Ph. Brüchmann, Die deutsche Marine (1848-1852), Archiv für Waffen- und Uniformkunde, No. 1, 1919, p. 26ff: Daggers as boarding weapons.
42 As long as thus produced.
43 Admiral Brommy's weapon also has this type of lower scabbard fitting, as an old photo shows.
44 German National Museum (ed.), 1848: Das Europa der Bilder—Michels März—Vol. II, p. 320ff.
45 To date two weapons of this type are known. The Schleswig Holstein naval officers' saber is very similar.
46 Lüder Arenhold, Vor 50 Jahren, Die Deutsche Reichsflotte 1848-1852, Berlin 1906, p. 2.
47 Lüder Arenhold, op. cit., p. 5; there "gold-decorated officers" are mentioned, so it can be assumed that they already wore a uniform.
48 This saber is in the Museum of Hamburg History (in the 1650-1860 shipping, harbor and city development hall, in 1995.
49 Josef Zienert, Die Schleswig-Holsteinische Marine 1848 bis 1851, Marine-Rundschau 1976, p. 227ff.
50 Die Schleswig-holsteinische Marine und ihre Leistungen während des dreijährigen Krieges vom Jahre 1848 bis 1851, circa 1851, in Beiheft zum Marine-Verordnungsblatt 1881, Vol. 33, p. 1, and Hans Hoffmann, Die Marine Schleswig-Holsteins, Köhlers Flottenkalender 1962, p. 88f.
51 Gerd Stolz to the author on January 8, 1995.
52 Gerd Stolz, Die Uniformierung der Schleswig-Holsteinischen Marine 1848-1852, Zeitschrift für Heereskunde 1986, p. 100f. [page 170]
53 The description may be based on the lithograph "Waffengattungen der schleswig-holsteinischen Armee" from "Up ewich ungeedelt: by Detlev von Liliencron, Hamburg 1898. [keep the hyphen: schleswig-holsteinischen]
54 In the literature this cap is often called the griffin head.
55 Bernhard Graser, Norddeutschlands Seemacht—Ihre Organisation, ihre Schiffe, ihre Häfen und ihre Bemannung, Leipzig 1870,

p. 480.
56 Heinrici, Die Norddeutsche Kriegsmarine, Berlin 1868, p. 18.
57 BA-MA, Signature RM 1/v771.
58 Rudolf Schuster (ed.), Deutsche Verfassungen, Munich 1978, p. 85f.
59 MVBl 1874, p. 28f.
60 Although it is not completely relevant, it may be noted that the 1872 uniform regulations the dagger is still mentioned as an interim weapon for sea officers. Since the 1874 uniform regulations no longer noted any chances to wear this weapon, it was dropped. Thus there is no official document that abolished the dagger.
61 MVBl 1875, p. 142.
62 More details are given in the Weapons of Honor section, including a description of Prince Adalbert's saber, which already bore an Imperial crown in October 1871.
63 An Army order that stated that spare parts for sidearms could be obtained only from the weapon factory in Erfurt as of January 1, 1878 (MWBl 1877, p. 222), may also have applied to the Navy.
64 Adalbert Mila, Uniformierungsliste des Deutschen Reichs-Heeres und der Kaiserlich Deutschen Marine, 3rd ed., Berlin 1876, p. 204.
65 When the model of a weapon had been prescribed definitely, it was "festgestellt".
66 Adalbert Mila, Uniformierungs-Liste des deutschen Reichs-Heeres und der Kaiserlich Deutschen Marine, 4th ed., Berlin 1881, p. 402.
67 Officers, doctors, engineers, deck officers, military officials received salaries. They had to supply their weapons themselves.
68 On ordnance weapons, see also Claus P. Stefanski, Ordonnanzwaffen!, Waffenfreund, No. 3, 1990, p. 4f.
69 Karl von Helldorf, Dienst-Vorschriften der Königlich Preussischen Armee, Part II, Berlin 1879, p. 10.
70 AVBl 1877, p. 222, no. 262.
71 MVBl 1884, p. 148, no. 122.
72 See also the 1883 writing cited above.
73 It must be noted that Prussian naval officers' sabers with fiscal stamps have not yet been found. Presumably service-prepared officers' weapons were not stamped before 1870.
74 See the thorough information in the work of Hans Reckendorff, Die Handwaffen der Königlich Preussischen und der Kaiserlichen Marine, Dortmund 1983, pp. 119-123.

75 MVBl 1889, p. 89, no. 109.
76 Hopmann, Das Logbuch eines deutschen Seeoffiziers, Berlin 1924, p. 80.
77 MVBl 1878, p. 1, no. 3.
78 From the Berlin Armeehaus catalog in Das Sponton, no place, 1965, No. 25-32, p. 67f.
79 MVBl 1888, p. 235, no. 185.
80 MVBl 1849, p. 137, no. 127, A.K.O. of June 12, 1849.
81 A similar model was used by Mecklenburg;s mounted infantry officers.
82 FWR = Friedericus Wilhelmus Rex (Friedrich Wilhelm IV); WR = Wilhelmus Rex (Wilhelm I).

83 Bekleidungs-Bestimmungen für die Seeoffiziere, Offiziere der Marineinfanterie, Maschinen- und Torpedo-Ingenieure, Feuerwerks\, Zeug- und Torpedooffiziere, Sanitätsoffiziere, Zahlmeister, Deckoffiziere usw., Seekadetten und Kadetten der Kaiserlichen Marine, sowie Bestimmungen über die Uniform der Seeoffiziere usw. des Beurlaubtenstandes, zur Disposition gestellter und verabschiedeter Offiziere usw. der Marine und Bestimmungen über das Tragen der verschiedenen Uniformen, Reprint of the 1890 issue, no place, 1893, p. XIII.
84 AVBl 1889, p. 78, no. 77.
85 The II. Sea Battalion was first established on April 1, 1889.
86 An exact date could not be determined.
87 AVBl 1905, p. 213, no. 156.
88 AVBl 1910, p. 323, no. 335.
89 Personal accounts from the I. and II. See Battalions show that a naval infantry battalion was set up in Kiauchau and was named the III. Sea Battalion after an Imperial order of June 13, 1898 (MVBl 1898, p. 181, no. 141).
90 MVBl 1900, Appendix to No. 3, p. I f., no. 1.
91 MVBl 1900, Appendix to no. 16, p. X, no. 8.
92 Rangliste der Kaiserlich Deutschen Land- und Seestreitkräfte für Ostasien (as of July 25, 1900), Burg b. M., no date.
93 Eberhard Hettler, Marineinfanterie und Schutztruppen, Kno:tel, Pietsch, Collas: 94
Das Deutsche Heer—Friedensuniformen bei Ausbruch des Weltkrieges, Stuttgart 1982, p. 354 and illustration, p. 351.
94 MVBl 1902, Appendiz to no. 11, p. V, no. 8.
95 MVBl 1900, Appendix to no. 16, p. X, no. 8.
96 As of 1903: Sabers for underveterinarians.
97 Leitfaden betreffend die Seitengewehre der Truppen zu Pferde und die Lanzen, Berlin 1891, p., 1 ff and Plate 4.
98 AVBl 1905, p. 213, no. 156.
99 AVBl 1910, p. 323, no. 335.
100 Zusammenstellung der Uniformen und Abzeichen der Beamten des Königlich Preussischen Heeres, Berlin 1902, p. 32, für Bekleidungsvorschrift für Offiziere,
Sanitätsoffiziere und Veterinäroffiziere des Königlich Preussischen Heeres
(O.Bkl.V.), May 15, 1899—Neuabdruck 1911, Berlin 1911, p. 155. [p. 172]
101 Verein deutccher Offiziere (ed.), Allgemeine Militär-Encyclopädie, Vol. 3, Leipzig 1869, p. 290.
102 Meyers Grosses Konversations-Lexikon, Vol. 5, no place, 1903, p. 416.
103 B. Poten, Handwörterbuch der gesamten Militärwissenschaft, Vol. 3, Bielefeld-Leipzig 1877, p. 77f,
104 See also Jörg Duppler, Prinz Adalbert von Preussen—Gründer der deutschen Marine, Herford 1986.
105 Adalbert Mila, Materialien zur Geschichte der Bekleidung und Ausrüstung der Kgl. Preuss. Armee seit 1805, Vol. 5, no place or date.
106 Reichskommissar Major Wissmann was also awarded a saber of honor by the Sultan of Zanzibar.
107 Meier, Erinnerungen an die Blockade Ostafrikas 1888/89, in M.O.V., April 1, 1931, No. 7, p. 159ff.
108 C. W. Allers, Unsere Marine (50 drawings), Breslau, no date (circa 1890).

109 In 1885 (according to A.K.O. of November 21, 1884) so-called Imperial prize weapons were already given as prizes in competition at the Military Shooting School. Extending the competition to the Army took place as of 1888. The weapons used by the troops were, unlike others, called prize-of-honor weapons. See also Claus P. Stefanski, Blankwaffen—Schiessauszeichnungen im Reichsheer, Zeitschrift für Heereskunde, No 31, 1987, p. 84ff.

110 MVBl 1903, p. 39, no. 34.

111 Gewehr-Schiessvorschrift für die Marine (Gw.S.V.), Berlin 1906, p. 95ff.

112 Olt Hüttman's saber is preserved in the Wehrgeschichtliche Museum in Rastatt (inv. no. 000568).

113 The shooting-prize weapons in the Army can be divided into prize weapons with acceptance stamps, and Imperial prize weapons, which could be won only by the Infantry Shooting School, but had no acceptance stamps, as did the naval prize weapons.

114 H. Frobenius, Militär-Lexikon, Berlin 1901, p. 360.

115 Dr. Schickert, Die Militärärztlichen Bildungsanstalten von ihrer Gründung bis zur Gegenwartm Berlin 1895, p. 166.

116 Bekleidungsbestimmungen für die Offiziere der Marineinfanterie, Berlin 1898, p. 44.

117 Bekleidungsvorschrift für Offiziere, Sanitätsoffiziere und Veterinäroffiziere des Königlich Preussischen Heeres (O.Bkl.V.) Of May 15, 1899—Neuabdruck 1911, Berlin 1911, p. 42.

118 "A la suite", as an honor, higher commanders or princely persons were often assigned to a troop unit; the designation also applied to officers who had long been furloughed and drew no salary.

119 Rangliste der Kaiserlich Deutschen Marine für das Jahr 1914, Berlin 1914, p. 6.

120 Nachrichtenblatt der Offiziervereinigung der Marine-Infanterie, 1963, No. 20, p. 18f.

121 K.M.V. of November 7, 1913.

122 Kgl. Sächsisches Militär-Verordnungsblatt 1913, p. 288, no. 253.

123 Kgl. Bayerische K.M.V. of December 9, 1913, no. 32459.

124 Württemberg's agreement could not be documented but was almost certainly given.

125 K.M.—No. 1377/11/14/A2—of 11/25/1914. [page 173]

126 K.M.—No. 2810/12.14.KRA—of 12/10/1914.

127 The first metal-saving order was followed by February 1916 by four other basic orders, as well as a series of further directives that dealt with special aspects of the subject.

128 K.M.: Armee-Verwaltungs-Departement—No. 2502.9.15.B3—of 10/4/1915.

129 K.M.—No. 1659/6.15/A2/—of 7/19/1915.

130 K.M.—No. 2752/12.16.A2—of 1/8/1917.

131 F. O. Busch & G. Ramlow, Traditionshandbuch der Kriegsmarine, Munich-Berlin, no date, p. 233 ff.

132 Marine-Archiv (ed.), Der Krieg zur See 1914-1918, Die Kämpfe der Kaiserlichen Marine in den Deutschen Kolonien, Part I, Tsingtau, Berlin 1935, p. 88.
133 Ernst Graf zu Reventlow et al., Unser Kaiser—25 Jahre der Regierung Kaiser Wilhelms II. 1888-1913, Berlin-Leipzig 1913, p. 237.
134 MVBl 1919, p. 325ff, no. 264.
135 MVBl 1919, p. 522, no. 411.
136 MVBl 1921, p. 203ff, no. 142.
137 MVBl 1921, p. 540ff, no. 475.
138 MVBl 1923, p. 1623ff, no. 216.
139 MVBl 1929, p. 86, no. 152.
140 St.O.Bef. No. 208, September 4-5, 1926.
141 Ostseestationsbefehl No. 15 of January 21, 1936.
142 RGBl I 1936, p. 145.
143 MVBl 1938, p. 561, no. 684.
144 MVBl 1925, p. 185, no. 252.
145 M.A. means Naval Artillery.
146 According to H.Dv. 240—Schiessvorschrift für Gewehr (Karabiner), leichtes Maschinengewehr und Pistole (Schiessv. f. Gew.) of June 30, 1934, Berlin 1937.
The authors could find no naval shooting rules for rifles etc. In the Katalog der Druckvorschriften der ehemaligen deutschen Wehrmacht—Part II Navy, Cuxhaven 1961—there are also no appropriate rules. So it seems that the Navy used the Army rules, which was also true in other cases, for example H.Dv. 242—Anschiessen der Handfeuerwaffen und MG (Erdwaffen), as M.Dv. No. 536 (analog).
147 J. Pype, Der Marinedolch, DWJ 1975, p. 540ff.
148 Lexikon des Kleidermachers, Vol. 3, Lepizig, no date (circa 1897), p. 76ff.
149 Bernd Diroll, Zur Geschichte der Hamburger SA, Militaria, No. 5 (Vol. 7), 5/13/1982.
150 Uniformen-Markt, No. 19, 10/1/1940, p. 146. [p. 174]
151 Old designation of a sea signal office.
152 Allgemeiner Marine-Befehl No. 31 of 1864, p. 33f.
153 Hanger, from the British designation of a dagger.
154 Ministerial-Blatt der Handels- und Gewerbe-Verwaltung 1903, No. 2, p. 11ff.
155 Nachrichtenblatt der Marineoffizier-Hilfe e.V.
156 Portepees are mentioned since 1705, see Constantin Kling, Geschichte der Bekleidung, Bewaffnung und Ausrüstung des Königlich Preussischen Heeres, Vol. 1, Die Infanterie, p. 231f.

157 Als alten Akten—Uniformierung der vom Königreich Preussischen Kriegsministerium im Jahre 1816 gegründeten Preussischen Marine, Beiheft vum MBVl 1876, p. 24f.

158 Paul Pietsch, Formations- und Uniformierungsgeschichte des preussischen Heeres 1808 bis 1914, Vol. I, Hamburg 1963, p. 132.

159 MWBl 1849, P. 96, no. 108.

160 Allgemeiner Marine-Befehl No. 31 of 1856, p. 103.

161 Gerd Stolz, Die Uniformierung der Schleswig-Holsteinischen Marine 1848-1852, Zeitschrift für Heereskunde, 1986, p. 100ff.

162 MVBl 1874, p. 1 and 28ff, no. 34.

163 MVBl 1921, P. 203ff, no. 142.

164 MVBl 1923, p. 162ff, no. 216.